Lewis A Tallerman

The Tallerman Treatment by the Local Application of Super-Heated Dry Air

Abstract of papers by medical men, reports from hospitals, and clinical demonstrations

Lewis A Tallerman

The Tallerman Treatment by the Local Application of Super-Heated Dry Air
*Abstract of papers by medical men, reports from hospitals, and clinical
demonstrations*

ISBN/EAN: 9783337038472

Printed in Europe, USA, Canada, Australia, Japan

Cover: Foto ©berggeist007 / pixelio.de

More available books at **www.hansebooks.com**

THE TALLERMAN
TREATMENT

BY THE

LOCAL APPLICATION OF SUPER-HEATED
DRY AIR.

ABSTRACT OF PAPERS BY MEDICAL MEN,
REPORTS FROM HOSPITALS,
AND CLINICAL DEMONSTRATIONS.

Reprinted from the Medical Journals of England, France,

Germany, the United States and Canada.

Institute:

50, WELBECK STREET, CAVENDISH SQUARE,

LONDON, W.

—

1898.

Principal Hospitals in the United Kingdom,
The Grand Duchy of Baden for the Government Establishment at Baden-Baden,
The Imperial Medical Clinic, University of Breslau.

Professor JAMES STEWART, at the Annual Meeting of the British Medical Association, September 1st, 1897.

In my opinion the most valuable of all methods of treatment is the use of baths of superheated dry air after the Tallerman method. It has been used in twenty cases of rheumatoid of arthritis in the Royal Victoria Hospital during the past nine months with gratifying results."—*British Medical Journal*, Oct. 30th, 1897

LAENNEC HOSPITAL, PARIS, Report and Cases.

' The Tallerman Treatment, on the one hand, has taken effect on diseases reputed to be incurable, such as chronic deformatory rheumatism, against which the physician until now confessed himself to be completely powerless, and on the other, it has acted more quickly and with better results than all other thereapeutic methods in diseases often intractable, such as sciatica, gout and urethral sinovitis."—*La Presse Médicale*, Jan. 26th, 1896.

NORTH-WEST LONDON HOSPITAL, Report and Cases.

" Tallerman's Treatment in Arthritis Deformans.—I must add I have never seen results so immediate and satisfactorily produced by any other treatment. It is now two years since this treatment was commenced in the first case, and yet the patient continues free from the complaint, and even the deformity to the fingers has greatly disappeared."—*Lancet*, 29th August, 1896.

Dr. MARTIN MENDELSSOHN (Demonstration by), at the Medical Society of Berlin.

" Mr. Tallerman, of London, has conceived the idea of employing superheated dry air locally applied. You see before you the apparatus which serves this purpose. A harmful reaction on the heart is avoided, so that the treatment is applicable even where cardiac weakness is present. Frequently, after a single good long sitting the joints are found to be relaxed and the movement freer, and in many instances the effect is perfectly astonishing—fixtures of the joints of long standing being so far improved as to restore the use of the limbs."—*Deutsche Medicinische Wochenschrift*, 17th March, 1898.

The "LANCET'S" recognition of the Inventor's attitude towards the Profession.

" Mr. Tallerman is to be greatly commended in that he has not endeavoured to bring his invention before the public, but has *confined its employment to the Medical Profession*. We feel sure that before long the Tallerman method of applying air at high temperatures to diseased portions of the body, *will take the place it deserves in the estimation of medical men, as the most satisfactory method at their disposal of treating many hitherto very intractable morbid conditions*."—*Lancet*, May 7th, 1898.

"THE MEDICAL PRESS," June 8th, 1898.

" It may safely be said that no one who undertakes the treatment of chronic diseases of the gouty rheumatic, or rheumatoid types, can expect to do full justice either to his patients or to himself, unless he has studied the claims of the Tallerman Treatment

INTRODUCTION.

50, WELBECK STREET,

CAVENDISH SQUARE, LONDON, W.

October 4th, 1898.

IN placing these abstracts before the Medical Profession, I beg to acknowledge the obligation I am under for the frank and unprejudiced expression of opinion upon the method of treatment associated with my name.

This, together with its reception in France and Germany and subsequent adoption by the Government of the Grand Duchy of Baden and the Imperial Medical Clinic of the University of Breslau, embodies the recognition of its merits by so large a number of eminent men as to fully confirm the title of the Tallerman Treatment to rank as a new remedy of the first importance, one which must have a far reaching effect and revolutionize the treatment of large classes of hitherto intractable and incurable diseases.

On that account it more than ever remains both my desire and ambition to promote its general use amongst the members of the Medical Profession, and I would gladly continue to use my influence with the proprietors to limit its employment to them, if a wish to that effect be practically manifested. It has already been so limited (See *Lancet*, May 7th, 1898) for the past five years, but not without considerable loss and sacrifice.

The contents hereof present a summary of the clinical and scientific investigations to which this new departure in therapeutics has been subjected, and to the results obtained therefrom is due that high reputation for efficiency and *absolute safety* it now enjoys. The evidence, derived as it is from sources of such an authoritative and representative a character, proves to a demonstration that patients can now be spared the suffering and expense of long journeys in search of relief and that they may be treated in their own homes, under the care of their own medical men, with a success unattainable by drug, electric, hydro-thermal or other remedies, whether used singly or in combination* in all such cases as :—Rheumatoid Arthritis, all forms of Gout and Rheumatism, whether Acute, Sub-acute, Chronic or Gonorrhœal, in Stiff and Painful Joints, Flat-foot, Sprain, Sciatica, Lumbago, Neuralgia, Neuritis, Writer's Cramp, Local Paralysis, Anæmia, Eczema, Chronic Ulcer and Bronchitis, Lead Poisoning, Dupuytren's Contraction, and for the relief of pain in certain uterine cases and in the breaking down of adhesions.

The therapeutic effects of the treatment—of which not the least important are :—

The immediate relief of pain and congestion,

The strengthening of the heart's action,

The elimination of morbid and acid products, and

The supply of nutrition to the whole system,

indicate its use in the practices of all medical men, and in the belief that it cannot fail to be generally adopted ; arrangements are being made to supply the apparatus at a *cost and on terms of payment* which will bring it *within the reach of every member* of the profession.

Attention is called to the improvements in the apparatus which have Improvements. not only extended its sphere of usefulness but by rendering it available for Inhalation. the *internal* as well as external application of dry air heated to any desired temperature, a PERFECT AND COMPLETE SYSTEM OF SUPERHEATED DRY-AIR TREATMENT IS PRESENTED.

* See Report by Professor Landouzy, page 14. and others.

Treatment by inhalation has not, up to the present time, been attended with satisfactory results, mainly owing to the difficulty experienced in volatilising drugs and substances requiring a very high temperature, and to the still greater difficulty of obtaining the entrance of these medicaments into the lower respiratory tracks and cells.

This difficulty has now been overcome by the adoption of certain fittings (supplied under my patent of 1896), by means of which the apparatus may be instantly made available for the internal application of superheated dry air by *inhalation* or *injection*, the vaporization being carried out before or during the process of inhalation at temperatures of from 150° to 200° F., and so on until the maximum which the patient can tolerate is reached.

Treatment without assistance? A further improvement and one which will be invaluable under certain circumstances is that by which patients (unless crippled in all their limbs) are enabled to treat themselves without assistance.

Safety. The Tallerman Apparatus has been designed and constructed to enable the use of any heating agent, and especially to provide for the perfect safety of patients under treatment, hence *there has not been a single accident* to record during the years it has been in existence.

Electric heating It is well known that the Company has had its electrically heated apparatus for the past four years, and now that the use of electricity has become more general, electric heaters specially constructed with the same regard for the safety of patients under treatment, will be supplied when required as well as the usual gas burners or oil lamp.

Gas heating. It might be explained that under the Company's arrangement for heating by gas, the necessary supply is taken from the gasalier or bracket in ordinary use at patients' houses, and that with the exercise of the most ordinary care in seeing that the vents are open, and that the gas is not turned up too high, the presence of an apparatus in operation cannot be detected by smoke or smell.

In conclusion, I would say that if I claim the support of the Medical profession, for the Tallerman Treatment and Apparatus, it is by reason of their originality and great therapeutic value. The unqualified recognition of both by men in the foremost ranks of medical scientists—and for no less a reason than that both apparatus and treatment have always been at the disposal of the profession, and have been worked throughout solely in connection with its members and their interests, notwithstanding the financial sacrifices it entailed, and the most tempting offers to act independently of them.

LEWIS A. TALLERMAN.

Communications should be addressed to *The Secretary*,

TALLERMAN TREATMENT INSTITUTE,

50, WELBECK STREET,

CAVENDISH SQUARE, LONDON, W.

Telegraphic Address—"TALLERMAN, LONDON."

Fuller details, reports and photographs of cases, will be found in " The Tallerman Treatment, " published by BAILLIERE, TINDALL, & COX, King William Street, London, W.

THE TALLERMAN APPARATUS.

PATENT.

FOR THE LOCALIZED APPLICATION OF SUPERHEATED DRY AIR (MEDICATED OR OTHERWISE), <u>EXTERNALLY</u> AS WELL AS <u>INTERNALLY.</u>

---

1. The apparatus IS PORTABLE.

2 . It can be heated by GAS, ELECTRICITY or OIL.

3. It can be TAKEN TO THE BEDSIDE OF PATIENTS and USED IN HOUSES HOWEVER REMOTE with ANY HEATING AGENT AVAILABLE.

4. Patients are enabled to TREAT THEMSELVES WITHOUT ASSISTANCE.

5. The apparatus may be converted into a HOT DRY-AIR INHALER, enabling the use of drugs requiring very high temperatures for volatilization.

6. It may likewise be rendered available for the INJECTION of heated air, medicated or otherwise, into wounds or passages of the body, and for the purpose of promoting artificial respiration.

The public Hospitals in which the "Tallerman Apparatus" has been tested and approved include the following:—

St. Bartholomew's Hospital	⎫
Charing Cross Hospital	
North-West London Hospital	
Children's Gt. Ormond Street Hospital	
St. Mary's Hospital	⎬ LONDON.
Metropolitan Hospital	
The London Hospital	
Guy's Hospital	
King's College Hospital	
St. Peter's Home, Kilburn	⎭
Salpétrière...	PARIS.
Laennec	PARIS.
Charité	BERLIN.
Augusta Hospital...	BERLIN.
Koenig Wilhelm Heilanstall	WIESBADEN.
Imperial Med. Klinik University ...	BRESLAU.
Hospital for Ruptured and Crippled	NEW YORK.
Philadelphia Hospital	PHILADELPHIA.
Royal Victoria Hospital	MONTREAL.
Dorset County Hospital	DORCHESTER.
Royal Portsmouth Hospital	PORTSMOUTH.
Sussex County Hospital	BRIGHTON.
Infirmary	CHELTENHAM.
Haslar Infirmary	HASLAR.
Union Hospital	CORK.
Norwich Hospital	NORWICH.
Sunderland Hospital	SUNDERLAND.
Children's and Women's Hospital ...	BRISTOL.
The General Hospital	BRISTOL.
Infirmary	LIVERPOOL.
North-West Derby Union	LIVERPOOL.
Livingstone Cottage Hospital... ...	DARTFORD.
Attenbrook Hospital	CAMBRIDGE.
Royal Infirmary	EDINBURGH.

The Treatment has been publicly demonstrated before members of the Medical Profession, and Papers have been read by them upon it at the following Medical Congresses and Meetings:—

British Medical Association, Annual Meeting, 1897.
British Medical Association (Scotch Branches), Feb., 1896.
German Medical Congress, Wiesbaden, 1898.
Society of Neurologists, Baden-Baden, 1898.
Philadelphia County Medical Society, November, 1896.
Society of Medical Practitioners, New York, 1896.
Royal Victoria Hospital to Medical Profession, Montreal, December, 1896,
Medical Society of Berlin, 1898.
Harveian Society, May, June, 1896, 1897 ; October, 1897.
North-West London Clinical Society.
Bristol Medico-Chirurgical Society.
And other Medical Meetings of a local character.

The Reports and Notices here reproduced are taken from the following Medical Journals :—

THE LANCET.
BRITISH MEDICAL JOURNAL.
CLINICAL JOURNAL.
MEDICAL TIMES.
MEDICAL PRESS.
MEDICAL MAGAZINE.
JOURNAL OF STATE MEDICINE.
PROVINCIAL MEDICAL JOURNAL.
PHARMACEUTICAL JOURNAL.
TREATMENT.
THE HOSPITAL.
WEST LONDON MEDICO-CHIRURGICAL JOURNAL.
DUBLIN JOURNAL OF MEDICAL SCIENCE.
BRISTOL MEDICO-CHIRURGICAL JOURNAL.
MONTHLY HOMŒOPATHIC REVIEW.
LA PRESSE MÉDICALE.
DEUTSCHE MEDICINISCHE WOCHENSCHRIFT.
MUNCHENER MEDICINISCHE WOCHENSCHRIFT.
WIENER KLINISCHE WOCHENSCHRIFT.
PROFESSOR VON LEYDEN'S "ZEITSCHRIFT FUR DIATETISCHE UND PHYSIKALISCHE THERAPIE."
PHILADELPHIA MEDICAL WORLD.
TRANSACTIONS AMERICAN ORTHOPEDIC ASSOCIATION.
PHILADELPHIA POLYCLINIC.
MEDICAL NEWS—NEW YORK.
MONTREAL MEDICAL JOURNAL.
WALSHAM'S DEFORMITIES OF THE FOOT.
WALSHAM'S THEORY AND PRACTICE OF SURGERY.
WALSHAM'S AMERICAN ORTHOPEDIC SOCIETY (PAPER TO). *See* TRANSACTIONS.

NOTICE.

The apparatus is patented, and each one bears a plate with the words "THE TALLERMAN TREATMENT PATENT," and its number. Notwithstanding this, its great success has led to the production of dangerous and cheap imitations, which have neither the endorsement of a single published hospital report as to their efficiency or safety, nor the recommendation of a single independent medical journal. If new apparatus are introduced claiming they are not imitations but produce the same effects as "Tallerman's," in a novel and different way then the same guarantees of the safety and efficiency of this new invention should be demanded in the shape of hospital notes and reports of official investigations as were demanded of and furnished by Mr. Tallerman, and are furnished here. If the apparatus and method of treatment are not novel (and heating by electricity is not), then to support any such imitation may be cheaper—stolen goods are supposed to be so—but the best interests of neither the public nor the profession can possibly be served by a policy as short-sighted as it would be wanting in appreciation. Meanwhile the Company think it only right to intimate that at the proper time it is intended to proceed against persons dealing with the said imitations.

Zeitschrift für Diätetische und Physikalische Therapie.—June, 1898.
EDITED BY
Professors E. VON LEYDEN and A. GOLDSCHEIDER.

ON THE

TALLERMAN TREATMENT

BY MEANS OF SUPERHEATED DRY AIR

AND

ON THE ACTION OF VERY HIGH TEMPERATURES UPON THE HUMAN ORGANISM IN HEALTH AND DISEASE.

By MARTIN MENDELSSOHN,

PRIVAT DOCENT IN THE UNIVERSITY OF BERLIN AND PHYSICIAN TO THE CHARITÉ HOSPITAL.

IT may be stated with a certain satisfaction that the medical art of our own day has witnessed a progressive development of the physical means of treatment, both in regard to increased knowledge of their action and more general recognition of their value. Among these means heat occupies a foremost position. An extraordinarily large number of methods of treatment, perhaps the majority of those belonging to the physical class, include the use of heat, and in every one of them the purely thermal action is, if not the only, at any rate by far the most real and efficient factor. It is true that mechanical and pharmaceutical effects also enter into nearly all these methods of applying heat; there are resins and salts and various other substances which assist in the action of medicinal baths, of warm and effervescing springs; there are chemical and physical influences which not only come into play in such applications as the steam, peat and fango baths, but include even electrical principles as in the sun baths, or purely mechanical effects as in the sand baths and similar methods of treatment. In all cases, however, the important and decisive element is the purely thermal action.

Now there is no doubt that in addition to the thermal effect all these accompanying elements exercise a varied and favourable influence in promoting the physiological end in view, so that this or that method of treatment appears better suited, according to the nature of its accompanying secondary elements, to a particular condition than some other allied method. But

B

on the other hand, it is equally certain that this very fact implies limitations and even disadvantages on the part of each particular method. In the first place unfavourable influences are caused by the technical peculiarities of the individual methods, and especially by the fact that some of them by their very nature are compelled to extend to the whole organism the action which is only required for a particular part. Thus they often produce powerful effects upon the breathing and circulation, which lie entirely outside the real therapeutic object, and appear as additional and undesirable symptoms. A second disadvantage attaching to some of them, such as the mud, peat and fango baths, is that they involve a process so unpleasant and dirty, that æsthetic considerations lead many susceptible patients either to abandon the treatment or to undergo it with disgust. Further, the fact that these methods can only be used in a particular place, and with particular appliances, is naturally incompatible with their wide-spread adoption.

Thermo therapeutics, therefore, as hitherto known and commonly used, are open to the following objections : first, that they are more or less accompanied by secondary and often objectionable influences ; secondly, that most of them involve a general re-action in the whole organism, which it is most desirable to exclude as far as possible ; and thirdly, that several of them present disadvantageous peculiarities in the inconvenience of their application and the uncleanliness of the vehicle used. They are also strictly limited both in regard to the degree of heat applied and the duration of the application. Any method of treatment, which is to bring the thermal factor to bear in the fullest measure, must therefore fulfil the following five conditions as far as possible :—

(1) The application of the heat must be as pure as possible —that is, free from simultaneous and accompanying effects of a different character ; the vehicle employed must itself be as indifferent as possible, and uncomplicated by the presence of other different substances or influences.

(2) Since all these processes only aim at effecting external action, whether local or general, the application must be so managed as actually to affect only the surface, and to leave the rest of the organism, and more particularly the air passages, untouched.

(3) The treatment must be, as far as possible, applicable everywhere, easy and convenient to handle, and agreeable, or at least not painful, to the patient.

(4) The temperature applied must be capable of being brought to bear at the highest possible degree of intensity.

(5) A single sitting must admit of the most extended duration possible.

Now I believe that all these conditions are fulfilled by a method, with which I have been carefully working for the last few months—the method of superheated dry air produced in London by Mr. L. A. Tallerman, but not hitherto known or tested in Germany. With the apparatus which Mr. Tallerman kindly placed at my disposal I have in the first place made extensive physiological observations on the action exercised upon the organism by this application of highly superheated air, and in the second I have carried out the systematic treatment of patients, and have recorded the therapeutic effects. I propose in what follows to state the results of my investigation. This of course involves the consideration of the subject from several points of view. It will first be necessary to give a description of the apparatus belonging to the treatment, and to explain the mode of applying it. Next, as a preliminary to further observations it will be necessary to prove a point, which had indeed to be tested at the outset—namely, that the application of such excessively high temperatures as 120° C and even 140° C is actually possible, and can be borne without ill effects or discomfort. I have already published a brief account of my conclusions on this head (April, 1898, Medical Congress, Wilsbaden) ; I shall here give the evidence on which they rest. This examination of the physiological action of high temperatures will be followed by the clinical portion of my subject, in which I shall discuss the treatment of affections indicated by the method and its effect upon them. After that, in order to fix the place of this new method of treatment among the other thermo-therapeutic methods, it will be necessary to examine the curative factors contained in the latter, the favourable and harmful by-influences involved, and also the extent and intensity of the thermal effect exercised by them. Finally, I shall discuss the physiological principles on which the curative action of these very high temperatures depends.

Tallerman method satisfies all the necessary conditions.

RE-ACTION OF THE LOCAL APPLICATION OF HIGH

TEMPERATURES UPON THE ACTION OF THE HEART, THE BODY

TEMPERATURE AND THE GENERAL CONDITION.

As has been already stated in the introductory observations, heat is the most important and effective factor in all thermo-therapeutic processes, so that its most intense and prolonged application must, *a priori*, produce the most powerful effects. The preliminary condition for such an extended use of heat for therapeutic purposes as is here contemplated is naturally this : that the application of these excessively high temperatures is not only possible, but can be accomplished without any subjective or objective ill-effects upon the patient. I must therefore show at the outset that such is in fact the case. I have tested the subjective and objective action of extremely high temperatures locally applied upon a very large number of persons, both

in health and in disease; and before entering upon the clinical
results and the dynamic value of this and allied methods of
treatment, I propose to set forth the physiological re-actions on
the general condition.

In the first place with regard to the subjective condition of
the patient during treatment, it is very little affected except
of course in respect of the sensation connected with general
perspiration and with the physical confinement in the apparatus.
It follows from the fact previously mentioned—namely, that an
uncomfortable or possibly burning sensation in the part under
treatment, is the only indicator required for having the
valves opened — it follows from this that the subjective
feeling in the limb under treatment, as I have repeatedly
convinced myself in my own person, is by no means disagreeable even at excessively high temperatures, at 140°C and more;
indeed it is less so than is usually the case in a very warm
atmosphere. The air inside feels warm, certainly, but by no
means particularly hot. What temperature can be borne and
for what length of time without any subjective discomfort may
be seen from the following observations :—

Extraordinary temperatures borne for three hours without discomfort in the Tallerman apparatus.

1.—E.R. Aged 26. Angina tonsillaris, otherwise healthy.
Right arm in the apparatus up to 1/3 of upper arm.

Time.		Temperature.		Sensation.
11.30		82° C		Arm introduced ; slight feeling of warmth
11.50	...	108° C	...	Moderate ditto, no burning
12.10	...	121° C	...	,, ,,
12.25	...	136° C	...	,, ,,
12.40	...	138° C	...	,, ,,
12.55	...	135° C	...	,, ,,
1.10	...	141° C	...	,, ,,
1.25	...	138° C	...	,, ,,
1.40	...	140° C	...	,, ,,
2. 0	...	139° C	...	,, ,,
2.25	...	141° C	...	taken out.

2.—A.T. Aged 63. Arterio sklerosis. Left leg to middle
of tibia introduced into the apparatus.

Time.		Temperature.		Sensation.
9.50	...	62° C	...	Leg introduced
10.10	...	96° C	...	Moderate feeling of warmth
10.30	...	118° C	...	Stronger ditto
10.50	...	121° C	...	,,
11.10	...	119° C	...	,,
11.30	...	123° C	...	Sensation of slight burning, valves opened for short time.
11.50	...	106° C	...	Moderate feeling of warmth.
12.10	...	118° C	...	,, ,,
12.30	...	117° C	...	,, ,,
12.50	...	121° C	...	,, ,,

Neither in these nor in other corresponding observations was any disagreeable sensation noticed apart from accidental and transient influences. Objectively also, the surface of the limb was found on being taken out to be moist and a little, but not excessively reddened ; in no other respect did it show any impression or change whatever, apart from increased movability of which more will be said later on.

A very far-reaching influence however is exercised by the action of the high temperature on the perspiration from the skin; primarily and in the highest degree from the limb treated but also from the whole surface of the body. In order to determine the amount of water lost I have repeatedly weighed the patient before and after treatment, and have ascertained that from 500 to 750 grammes of water are lost by the application of a medium temperature for a moderate length of time. *Far-reaching influence by the action on the perspiration.*

The loss of water thus effected through the skin in a comparatively short time doubtless takes place mainly from the surface of the limb treated. Up to the present I have not been able to make any exact observations on the relative share taken in the process by the portion of the body inside the apparatus and by that outside, though I may do so at some future time. The relation I have indicated is, however, clear enough from the visible effects and from a consideration of the condition. Moreover a moment's reflections shows the necessity of a copious prolonged and continuous perspiration from the surface of the limb treated, for it is obviously this constant perspiration alone which makes the excessively high temperature endurable and prevents any injury to the skin. As soon as ever the perspiration is checked discomfort ensues. Of course a raising of the temperature of the skin itself or still more of the underlying tissues as high as 100° C is absolutely unthinkable, but even the rise of only a moderate number of degrees would be intolerable. The organism protects itself here as elsewhere against high temperatures by producing a copious perspiration, which as it evaporates from the skin into the surrounding air lowers the temperature of the surface so as to neutralize the external heat. This is the simple condition which enables these excessive temperatures not only to be borne but to be unaccompanied by any special sensation of heat. And since the amount of evaporation depends not only upon the temperature and atmospheric pressure, but also in an important measure upon the degree of moisture present in the air into which the evaporation takes place, it is through the provision for keeping the air dry and as free as possible from moisture that we reach at once the possibility and the necessity of maintaining the evaporation of perspired moisture to a far greater extent and far more constantly than with any other existing method of applying heat. Further, the rate of this evaporating process must be relatively all the higher, when the total quantity of water, large in itself, that is drawn from the body—as proved by *Protection afforded by perspiration against excessive temperatures.*

the foregoing figures—comes from a relatively small portion of the surface, even from the hand only, the forearm and part of the upper arm.   For even if the whole body be induced, by being carefully wrapped up and by the simultaneously imbibition of water, to take part in the perspiration, the latter is nevertheless mainly a local process affecting the part of the body directly submitted to the action of heat.

The remarkable effect upon the whole circulation exercised by this increased determination of fluid to an isolated part of the body, and the reaction thereby set up in the movement of the lymph and the venous blood, I shall have occasion to discuss more fully later on.

That this neutralization of the enormous difference of temperature between the surface of the body and the surrounding air actually takes place at the skin and is no merely theoretical supposition, may be directly proved.   If a patient, whose arm was lying in the apparatus, held in his hand a thermometer, the bulb of which was kept the whole time between his fingers in contact with the skin, every time it was withdrawn this thermometer showed only trifling rises, the highest of which are given in the following table together with the corresponding temperature of the free air in the apparatus:—

Demonstration of the neutralizing effect upon the temperature.

3.—A. H., Aged 32.   Healthy.   Right arm introduced into the apparatus up to middle of the upper arm.

Time.			Temperature.			Internal Temperature.
11. 0	...	...	84° C	...	...	37·2°
12. 0	...	...	122° C	...	...	—
12.15	...	...	120° C	...	...	—
12.30	...	...	121° C	...	...	—
12 45	...	...	123° C	...	...	—
1. 0	...	...	121° C	...	...	—
1.15	...	...	120° C	...	...	38·7° C

4.—M.V.   Aged 39.   Gastric catarrh.   Left leg introduced a little above the knee.

Time.		Temperature.			Internal Temperature.
11.20	...	80° C	...	...	36·3° C
11.35	...	102° C	...	...	36·8° C
12. 5	...	124° C	...	...	37·0° C
12.25	...	120° C	...	...	37·1° C
12.45	...	122° C	...	...	37·0° C
1. 0	...	125° C	...	...	37·2° C
1.15	...	122° C	...	...	37·3° C

Whatever degree of heat is applied to the part treated the general temperature only rises a fraction of a degree centigrade, never more, whether it be taken under the tongue, in the axilla or in other places.

I add a further observation :

5.—A.D. Aged 41. Bronchial catarrh. Left arm introduced nearly to the shoulder.

Time.	Temperature.	Temp. in right axilla.
11. 0	... ... 65° C ...	... 36·6° C
11.15	... ... 90° C ...	.. 36·7° C
11.30	... ... 112° C ...	... 36·9° C
11.45	... ... 121° C ...	... 36·9° C
12. 0	... ... 124° C ...	... 37·1° C
12.15	... ... 128° C ...	... 37·0° C
12.30	... ... 126° C ...	... 37·3° C
12.45	... ... 128° C ...	... 37·3° C
12.50	... ... taken out ...	... 37·4° C
1. 0	Room Temperature 14·5° C ...	... 37·3° C
1.15	,, ,, ...	... 37·3° C
1.30	,, ,, ...	... 37·1° C
1.45	,, ,, ...	... 37·1° C

These figures show in agreement with general experience that at first the body temperature rises moderately compared with the air, but that degree of elevation never exceeds a moderate limit. On the other hand, after withdrawal, this slight elevation only subsides with relative slowness, and is almost always still plainly perceptible after the lapse of a full hour.

The effect on the heart, which is the most important point, is equally small. I shall take occasion later on to say something about the respective claims of particular thermo-therapeutic methods to influence the circulation, and can only state here that the Tallerman treatment at present in question, involves no reaction upon the heart and circulation worth mentioning. Out of the numerous observations two may be taken as typical :—

6.—S.S. Aged 18. Healthy. Right leg introduced to upper third of tibia.

Patients treated for three hours a 250° F. to prove the heart is not unfavourably affected.

Time.	Temperature.	Pulse.
9 30	60° C	106·
10. 0	108° C	82·
10.20	119° C	78·
10.40	120° C	78·
11. 0	118° C	80·
11.20	119° C	82·
11.40	121° C	82·
12. 0	119° C	84·
12.20	120° C	86·

7.—A.L. Aged 46. Bronchial asthma. Right arm introduced up to middle of upper arm.

Time.	Temperature.		Pulse.
9.30	90° C	...	86·
9 45	122° C	...	82·
10. 0	128° C	...	78·
10.15	132° C	...	80·
10.30	128° C	..	86·
10.45	130° C	...	88·
11. 0	131° C	...	92·
11.15	... taken out	...	96·
11.30	Room Temperature 14.8° C	...	90·
11.45	,, ,,	...	86·
12. 0	,, ,,	...	80·
12.15	,, ,,	...	76·

Here, too, we see that the action of the heart is only quickened gradually and slightly by a few beats, after it has returned to the normal from the initial excitement which is obviously due to mental causes. This quickening also returns to normal as soon as the operation is over ; only gradually, it is true, but in all cases that have come under observation steadily and without interruption.

The harmlessness of the treatment declared proved. The foregoing is sufficient to prove that the first condition for treatment by high temperatures—namely, the harmlessness of the method and its trifling reaction on the general health—is here fulfilled. Having first established that fact, I then went on to submit suitable patients to the treatment, and in what follows, I propose to give the results and the information obtained, concluding with an examination of the physiological significance of the curative effects, and a comparison between this and other allied but different therapeutic procedures.

ST. BARTHOLOMEW'S HOSPITAL.

CLINICAL LECTURE ON THE TALLERMAN TREATMENT.

(See "*Clinical Journal*," May 30th, 1894.)

A Clinical Lecture on the Therapeutic Action and Uses of the Localized Application of Dry Air heated to High Temperature in Certain Classes of Surgical Affections. Delivered at St. Bartholomew's Hospital, on Wednesday, May 23, 1894, by ALFRED WILLETT, F.R.C.S., Surgeon to the Hospital.

ON January 23rd, at the courteous invitation of Mr. Lewis A. Tallerman, I went to 1, Chiswell Street with Mr. Walsham and Mr. Paterson, my house-surgeon. There were some three or four other medical men present. Of the few patients collected two were selected for trial. The first was a middle-aged man with subacute synovitis of the knee. Increased heat, slight effusion, and some pain existed. This was increased on attempting to move the knee, which was held semi-flexed, having a range of active movement of only 10-15°. He could only walk, or rather, I should say. hobble, with the aid of two sticks, on the toes of the affected limb. In this condition he was placed in a cylinder like the one we have here. When taken out of the bath, after some thirty or forty minutes, the knee was straight, all pain had left him, the foot was on the ground. and he walked almost briskly out of the office. I heard that in a few days' time he returned to work. This result naturally made a considerable impression on me, for I know of no treatment that could have brought about so rapid a cure. The case, no doubt, was not only curable, but well on the road to recovery. At the time I said that in hospital I should have anæsthetized the patient, manipulated his knee, and then brought it straight ; but even so I feel sure that many days would have elapsed before he would have been well, whilst here the man had been apparently almost cured in little more than half an hour, not only quite painlessly. but by a process that one might almost call that of luxurious ease.

Subacute synovitis, successful treatment of.

The treatment one of luxurious ease.

The second case was equally satisfactory—that of a woman of about 45, with both hands crippled by chronic gout. The fingers were all kept slightly flexed. Before going into the bath she could neither straighten nor flex them. It appeared that slowly, for upwards of a year, she had been drifting into this condition, and now she could do but little for herself. Only one hand, the right—said to be the worse—was put into the bath. After fifteen minutes she volunteered the statement that her fingers seemed to loosen, and soon after she could oppose the thumb to all the fingers, which for many months she had not been able to do. After half an hour the hand was removed from the bath. I saw her open and close her hand readily, while the left, which had not been treated, remained in the same fixed position.

Chronic gout. successful treatment of.

Both these patients were entirely comfortable all through the process. The patients perspired freely, and when the limbs were taken out of the bath they were very red and moist.

Subsequently, I was asked if I would test the efficacy of the bath at this hospital, and having obtained the consent of the authorities, I have done so, and desire now to put the results before you.

I will now read you the abstract of the notes of cases which were treated in President and Pitcairn Wards, and comment on them in turn as occasion requires ; for time does not permit, neither is it necessary, that I should weary you with the full notes.

Rheumatoid
arthritis, the first
case of, treated
in the London
Hospitals.

Alice C., 25 years of age, was admitted into President Ward on September 29th. Her history was one of gradual onset of pains seven years previously, a dull aching by day and night. Six years ago, or one year after the onset of these pains, she had rheumatic fever. She was under treatment at Reading, and subsequently at Bath. On admission, her condition was that of an anæmic girl, with very pained expression ; heart's sounds natural. She was almost helpless and bedridden from crippled joints. The affected joints were the right ankle, knee, and hip, the left hip and both elbow and wrist joints ; all these were stiff and painful. The right hip and knee were flexed at angles respectively of 165° and 90° ; both elbows were flexed to about 120°, with about 35° angle of movement. The wrists were completely anchylosed. The changes chiefly affected the fibrous structures of the joints, but about the elbows, especially the left, were bony and cartilaginous defects, with creaking and grating on movement in both. The electrical reactions were normal.

On January 17th the right hip, knee, ankle and right elbow and wrist joints, were all forcibly broken down and moved under an anæsthetic. The right hip and knee were brought straight and fixed upon a Thomas's hip-joint splint. On the 25th, under passive movement there was some improvement. Nitrous oxide gas was again administered for wrenching the right wrist and elbow. At this time the left elbow was the worse, but I thought it better to confine myself to the joints on the right side of the patient's body. She was still being treated by passive movement and given cod-liver oil. On February 20th, Tallerman's hot air bath was used at a temperature of 260° for fifty minutes.* There was much less pain afterwards ; the range of motion was *not increased ; the movements, however, were more free through the given range. But on any stretching of the joints, with the view to obtaining increased range of movement, the old pain was felt. The baths were continued from February 20th to March 21st. Up to this time there was no very definite improvement beyond the above changes. On April 9th gas was administered, and the left elbow extended and flexed, all the adhesions in it being broken down, and afterwards placed in the bath. On the 19th it was noticed there was general improvement in all her joints. The left elbow, which was the worse, was much less painful and more mobile.

I now show you the patient, who, as I mentioned, was admitted practically bedridden and helpless just at the end of last year. She left the hospital walking on crutches. Any movement was attended with great pain when she came here ; on leaving, the joint movements were much freer, almost painless, and the patient was able to do almost everything for herself. These noticeable changes, relief from pain and freer

---

* Used throughout on left arm and hand only.

movement, were mainly attributable, I think, to the influence of the hot-air baths. She was under the bath treatment for about eight weeks. On asking her to strike at my hand, we find that the right upper extremity is the stronger, but with both she can hit my hand a fairly vigorous blow; on admission it would have been absolutely impossible for her to have attempted anything of the kind. I mentioned the great creaking which occurred in the joints; now within the range of movement the joint has become quite natural. The range of movement in her left elbow-joint is better than it is in the right; the flexion is a little greater, the extension perhaps not quite so good, but here again the movement, although not absolutely smooth, or so smooth as in the right, is accompanied by less marked creaking than when admitted. The wrist-joints were absolutely fixed. She has a moderate amount of pronation and supination; still if I put any strain upon the joints it is obvious that it pains her. Her right wrist remains very stiff.

The next case was that of a man named James L., 26 years of age, a plumber, suffering from urethral synovitis. One month he has had discharge, for two weeks and a half pain in the left knee, afterwards in the right ankle and both feet. On admission his left knee, right ankle and both feet were swollen, tender and hot, but there was no marked redness. There was no cardiac murmur and no sweating. The temperature was raised 2-3°. Past history : Rheumatic fever twenty years ago. During the first two weeks after admission he had much pain in the knees, ankles and feet continuously, but varying in intensity. His joints were all stiff. His temperature was raised usually, frequently to 101°. He was relieved by blisters. The movements of the joints improved.

About February 20th he was placed in the hot air baths. From that time his improvement commenced. He left the hospital at the end of March. The movements of his joints were greatly improved, and all active inflammatory condition passed away. His knees and ankles and right shoulder were slightly painful but the movements were nearly natural. *This was an extremely severe case of that form of rheumatism, and one which, I think, might have left him crippled by adhesions,* although he would have got well in time ; yet *no improvement took place until one ankle and knee were placed in the cylinder, and from that time convalescence commenced. I feel convinced the bath treatment greatly accelerated recovery and probably prevented some serious organic changes in the joints. That patient was five weeks under treatment.*

The next was a man, named Benjamin D., 22, a potman, the subject of talipes planus. He had suffered from rheumatic gout, three months previously in the ankles. From that time he had pain on standing. For the two months before admission he had less pain but great weakness of the ankles. It was thought that this case would be benefited by treatment in the

Gonorrhœal rheumatism remarkable success in.

cylinder; here the bath failed to benefit beyond affording relief to pain and stiffness whilst in the bath. The result, perhaps, might have been different if this patient had at the time been an in-patient, but being an out-patient, and having to go away and walk about prejudiced the chances of recovery. Still the case was only treated with the baths as a test case.

Cecilia L., aged 17, was admitted into President Ward on January 16th, suffering from painful knee, believed to be a subacute tubercular affection. In June, 1893, she had been in President Ward under Mr. Cripps, and the limb was then brought into a good position and put up in plaster of Paris. Some enlarged cervical glands at that time were also removed, and hence there is good reason for believing that the diagnosis of tubercular affection was correct. In the meantime, she had been from time to time treated by reapplication of plaster of Paris splints, and once leather splints were moulded to her knee, and worn for a time. Recently there had been an exacerbation of pain and swelling, and for this she was readmitted. She was of the rather dull, sleepy type of patient. There was some backward displacement of the tibia, with a tendency to rotation of the tibia inwards rather than outwards. The patient usually lay, not on the affected side, but upon the opposite side and rolled her limb inwards. This, of course, kept the crucial ligaments tight, and also prevented backward and outward displacement. On the 17th of January, on account of localized pain an icebag was applied; and in February a ten-pound weight was adjusted, position remaining good. On February 3rd the limb was again put into plaster of Paris. It was taken out on the 9th. On the 20th the hot-air bath at a temperature of 170° F. was used for half an hour. This did not seem to increase the movement, but did seem to diminish the pain. Ultimately she left the hospital wearing a Thomas's knee splint. The leg was straight and firm and much less painful, and her general state of health was improved. This patient had the bath on only two or three occasions. Probably the affection was rather too acute, or, if not so, the patient was of that neurotic type in whom it is difficult to tell whether the pain is real or imagined. But it did not, on the whole, seem a very fair test case for the bath treatment.

*   *   *   *

Temperature 300° F., checked by inside thermo-meter level with the patient's hand.

At one séance in this hospital the thermometer was withdrawn by one of the surgeons, Mr. Walsham, and found to register 300°. The thermometer had been placed inside to check the one outside the cylinder, and was on a level with the patient's hand.

Experience with treatment distinctly encouraging.

In offering some comments on this invention I must say at once that I can only express a very guarded opinion at present, mainly because my experience is still very small; yet small as it is, I have no hesitation in saying it is distinctly encouraging.

In the cases that have been treated in President and Pitcairn Wards I was anxious to try the bath in an assortment of diseases, and also to test its action purely and simply. In the next place, the process occupies roughly an hour for each patient, and some of the patients were undergoing the bath treatment for the entire two months. Mr. Tallerman kindly gave his valuable services in supervising its administration; hence, during the two available hours of the afternoon with two baths going it was not possible, as a rule, to treat more than about four patients a day, and of these some were under Mr. Walsham's care.

* * *

We note that the part, foot or hand, issues from the bath very much the colour of a boiled lobster; the flow of blood in the skin is obviously, and that in the subcutaneous tissues probably, increased greatly. The third effect noted is what we may term its anodyne influence. In most cases, not merely is pain relieved, but often it will be entirely removed. This is shown in many ways. The patients invariably experience relief. Then they will use the limbs with much greater freedom. Movements that excited pain before can be performed after the limb is placed in the bath without pain. Again, in cases of breaking down of adhesions under gas the patient is in great pain afterwards; but let the limb be placed in the heated cylinder and the pain is greatly lessened, as also is the secondary stiffness, due to extravasation and inflammation round the torn tissues. On the other hand, there is none of the excitement, amounting in some individuals to distress, from breathing the hot air of the Turkish bath before free perspiration breaks out. The patient with arm or leg in this cylinder seems throughout in the most absolute comfort.

Effects of the treatment. Greatly increased circulation. Pain relieved or entirely removed, Greater freedom of movement. Value where adhesions are broken down. Absolute comfort of patient throughout.

If such are the results observed from the baths in healthy persons, its therapeutic influence in certain affections is obvious, and, I think, clear to demonstration. The next consideration, therefore, is, To what class of disease is this plan of treatment likely to be serviceable? I can only speak of its use in surgical affections. Of these, sprains, stiff joints—those where movement is more or less limited—callous limbs, flat-foot, gonorrhœal rheumatism, and possibly some skin affections, are the most likely to be benefited. The cases I have recorded show a fair selection of these.

This treatment serviceable in various diseases— Sprains, Stiff joints, Callous limbs, Flat foot, Gonorrhœal rheumatism, Skin affections.

On the third point, as to its being an adjunct to surgical treatment, I have only to repeat what I have already said, as to the great assistance this hot bath affords after breaking down articular adhesions, both in relieving pain and lessening the tendency to recurring stiffening. I think that, in addition to these, the bath may be occasionally useful as an adjunct to the electrical treatment of certain paralytic cases, and generally useful, I think, before massage to wasted muscles.

Treatment of great assistance to surgeons.

Medical cases in which the treatment should be valuable, as Rheumatic arthritis, Gout attacks, Sciatica, Lumbago, Spinal cases.

I think there is evidence that the effect of the bath is not confined solely to the part acted on, for the temperature of the patient is raised usually nearly one degree; true, this elevation of temperature alone would not prove anything, for excitement will often send up temperatures; but the entire skin becomes relaxed and perspiration occurs freely over the body. Besides, patients unite in saying that with subsidence of pain in the part treated pain is lulled in other joints. *I hope some physician will be induced to try the hot air system in selected medical cases. By analogy, I think, that in rheumatic arthritis, gouty attacks in arms cr legs, sciatica, lumbago, and perhaps in some spinal cases, good results might be anticipated.*

The adoption of the treatment at the hospital recommended strongly.

In so many surgical cases should this hot-air bath treatment prove serviceable that I hope it will not be long ere one is available at any time in this hospital, and also that we have a nurse trained to its use.

Report by Prof. LANDOUZY of the Laennec Hospital, Paris.

August 3rd, 1897.

From what I have seen of the local application of hot air by the Tallerman method and apparatus, both in my wards at the Laennec Hospital and in my private practice, I am enabled to say :

A real advance.

1. That the Tallerman method constitutes a real advance in the local applications of hot air.

Better than other remedies in many acute, sub-acute, and chronic affections, and particularly so in— Chronic rheumatism, Gonorrhœal rheumatism, Acute gout, Sciatica.

2. That this treatment is indicated in a large number of acute, and a still larger number of subacute and chronic affections, both of the limbs and the adjoining parts. In these it seems to produce better results than the remedies hitherto in use. In particular, cases of gonorrhœal rheumatism, of joint affection in chronic rheumatism, of acute gout and sciatica, have been better and more rapidly relieved, better and more rapidly reduced, than by the old external remedies, whether used alone or in combination with drugs.

3. That the indications and contra-indications of its use require more precise definition, in order to determine the important place which the treatment deserves to take in the therapeutics of those painful affections of the joints which are so common in the course of infectious and constitutional diseases.

Report by Prof. DEJÉRINE of the Salpêtrière Hospital, Paris.

August 7th, 1897.

Most remarkable results.

Among the cases of chronic rheumatism in my wards at the Salpêtrière, in which I have been able to appreciate good effects of the high-temperature treatment, I have observed the following, in which this method has given the most remarkable results.

The patient was a young woman, 25 years of age, suffering from unilateral infectious rheumatism of the left side, the commencement of which dated from October, 1896. I saw her for the first time four months after the beginning of the illness, and it was the most severe case that I have ever seen. The elbow, wrist, knee, and ankle-joints were all greatly swollen, the sheaths of the tendons and serous brosæ much enlarged and excessively painful. She had been unable to leave her bed for four months and a half. On coming to Paris at the beginning of April she was placed under daily treatment by the cylinder at 212°, 230°, and 240°. At the end of a few sittings the pains had almost disappeared, and the swelling began to subside. At the present time, August 7, 1897, after 62 sittings, the patient's condition is greatly improved, and the joints have returned to their normal size.

I consider, therefore, that in this case the results of the treatment with superheated dry air by the Tallerman method are most remarkable, and superior to everything else that I have seen used in similar cases. *(Superior to everything else.)*

THE LAENNEC HOSPITAL, PARIS.

Report and Notes of Cases treated under Prof. LANDOUZY and Dr. OULMONT. Paper by Dr. EDOUARD CHRETIEN, of the Laennec and Salpêtrière Hospitals, Paris.

La Presse Médicale, 26th December, 1896.

Paper by Dr. EDOUARD CHRÉTIEN, of the Laennec and Salpêtrière Hospitals, Paris.

IN December, 1895, Mr. Lewis A. Tallerman demonstrated in the clinic of M. Oulmont, with which I had the honour to be connected at that time, his method of treatment and the effect produced by his apparatus. *(Demonstration at the Laennec.)*

* * * *

With the Tallerman apparatus, thanks to its peculiar construction, we can apply hot dry air to large portions of the body—a whole limb, for example—and can attain temperatures in the neighbourhood of 300° F., or 141° C., or even higher. Observation has shown that in cases where several joints are affected, the hot-air bath acts not only on the joint enclosed in the apparatus, its influence is equally felt by the other affected joints, even those on the other side of the body. The effect obtained is only less marked. So when, for one reason or another, it is impossible to treat the affected part directly, we *(Limbs treated indirectly with success.)*

have tried, and successfully, too, to treat it indirectly, by placing the corresponding sound limb in the apparatus. The effect is not so marked, I repeat, but it makes a decided impression nevertheless.

*   *   *   *

Increase in the number of cases which will benefit by the treatment. When Mr. Tallerman placed his apparatus in our hands and asked us to test it, he declared that he had employed it in England to treat a great variety of diseases, such as rheumatic arthritis, gout, sprains, acute joint affections, stiff joints resulting from injuries or inflammations, gonorrhoeal rheumatism, flat-foot, chronic and deformatory rheumatism. Since then the list of diseases which it will benefit has been greatly added to, and we can see still more which it may benefit.

Among them are muscular rheumatism, acute synovitis, sciatica, lumbago, tuberculosis, arthritis, certain forms of neuralgia, peripheral neuritis, etc.

*   *   *   *

Professor Landouzy and Dejérine anxious to authorise employment of apparatus in Laennec and Salpêtrière. My chiefs (*maîtres*), Professors Landouzy and Dejérine, to whom I had spoken of the apparatus, were equally anxious to authorize its employment in their clinics in the hospitals of Laennec and Salpetrière.

I will now describe briefly the results obtained in Paris with a large variety of patients, both in the hospitals and in private practice, and will follow that with some observations and reflections on this treatment, what it has done, and the part it is destined to play.

The results obtained remarkable enough to merit publication. Thus one can judge by these few cases described that the effects obtained by the Tallerman apparatus in chronic deformatory rheumatism, gonorrhoeal rheumatism, sprain, sciatica, gout, and certain forms of joint disease of a more or less well-known origin, have been remarkable enough to merit publication. I speak only of these diseases because these are the only ones which we have treated in Paris, but the list of diseases treated in England is, as we have seen, much longer.

Disappearance of pain. The most evident result from this treatment, both to patient and physician, is the disappearance of pain.

What may be hoped for from the treatment. What can we hope for from the superheated dry-air treatment ? What are the favourable and unfavourable symptoms produced by it ? There are no unfavourable symptoms that I know of resulting from the application of the treatment. Thus, Improvement of the heart in case No. 7, under treatment for rheumatism. as was proved by case No. 7, the presence of a heart trouble did not forbid the use of the Tallerman treatment ; quite the contrary. In fact, the diminution of the arterial tension caused by the peripheral dilatation of the bloodvessels facilitated the heart action, as was indicated by the immediate acceleration of the pulse.

As far as affections of the respiratory organs are concerned, I have no personal experience. All that I can say is, that in several of the cases published in the English medical journals, the patients who presented themselves for treatment had, in addition to their joint affections, chronic bronchitis. That was so much helped by a course of these superheated dry-air baths that the treatment has been deliberately applied in cases of affections of the respiratory organs, and with success.

There remain still the effects on the renal organs. The English physicians are mute on this subject. I have carefully examined the urine of all the patients who were treated in the Laennec Hospital and found no change either in quantity or quality (sugar, albumen). That is a point which has not, up till now, been cleared up.

A priori, it would not seem that the hot-air baths could have any evil influence on the kidneys in their normal condition. It would be a benefit, on the contrary, in certain nephritic cases if they would, by diminishing the muscular tension, diminish the diuresis. *Beneficial influence on the kidneys.*

The analysis of urine has not only been directed to the discovery of the modifications which could have existed before the treatment: we have also sought to discover if the hot-air baths influenced the quantity of matter secreted by the urinary organs during the twenty-four hours. *Analysis of urine, great increase in the quantity of uric acid eliminated.*

In case No. 5 the urine was analyzed after each bath. It showed only a very slight improvement in the elimination of salts, particularly chlorides. The daily co-efficient of the urine changed from 20·97 grains to 25·50 grains.

From this point of view, case No. 12 is most interesting, for here, in a case of long-standing gout, the daily quantity of uric acid eliminated rose from 57 centigrammes after the fourth bath to 89 centigrammes after the ninth. The patient being still under treatment, we cannot say now how far this elimination of uric acid will go, and when it will return to normal.

The field of diseases suitable for treatment by these super-heated dry-air baths is, as one can see, very large. It comprises: *Very large field of diseases suitable for treatment; will do what no therapeutic agent up till now has accomplished.*

1. All the painful affections of the limbs, the diseases which attack the joints, the diseases of the muscles, the trunk nerves (sciatica, neuritis), and perhaps we can use it in the treatment of wasting diseases, diseases of the bone and of syphilitic origin.

2. Acute and chronic arthritic diseases, whether acquired (gout, rheumatic fever, gonorrhœal, tuberculous) or hereditary. I do not wish to say that this treatment can cure the deformities of old sufferers from rheumatism, their fibrous adhesions or their muscular atrophies. But I believe that it will abate them, and after treatment with the hot-air baths we can with much greater ease and with better results break the fibrous adhesions

C

CASE 8.—April 7th, 1896, before treatment (see also skiagraph of right knee):
showing "swollen and crippled joints, hip joints stiff and painful, marked
"Cardosis and upward tilting of pelvis, cannot kneel or walk upstairs, cannot
"separate knees, cannot close hands or raise them higher than chin, wrists
"enlarged, stiff and painful, can only walk a few steps with evident pain,
"neurotic, restless at night, and sometimes faints.

For Case notes and Report, see page 24, " The Tallerman Treatment "
(Baillière, Tindall, & Cox, London).

CASE 8.—June 10th, 1896, after treatment (see also skiagraph of right knee): Patient able to kneel, cross legs and walk up and down stairs, partially flex fingers on palms and raise the arms above the head, joints painless on forcible movement, patient much improved in her general health.

After above patient treated a few times after October, 1896: *Medical Report,* 28th July, 1897—"Patient grown and looks well, can execute ordinary "movements as well as any chi'd her age, can button boots, pick up pins "from carpet, kneel, dress and undress herself, do needlework, standing on "floor can raise feet on to ordinary chair."

For Case notes and Report, see page 24, "The Tallerman Treatment" (Baillière, Tindall, & Cox London).

CASE 8.—Skiagraph of right knee before treatment: effusion into the joint, only a few degrees of flexion at the expense of much pain.

For Case notes and Report, see page 24, "The Tallerman Treatment"
(Baillière Tindall, & Cox, London).

CASE 8.—Skiagraph of right knee after treatment : normal position of patella.

For Case notes and Report, see page 24, "The Tallerman Treatment
(Baillière, Tindall, & Cox, London).

and restore movement to the stiff joints. I believe that by associating the hot-air treatment, electricity, and massage, we can, if not cure, at least considerably ameliorate the lot of these sufferers whom we see in the chronic disease wards, helpless and deformed from rheumatism for ten, fifteen and twenty years, and restore again the charm to their life, something which no therapeutic agent employed up till now has been able to accomplish.

Used in tuberculous cases.

As far as the joint affections are concerned, and more particularly in tuberculous cases, it is possible that the high temperature obtained by the Tallerman treatment will give us good results. In fact, the Koch bacillus offers a poor resistance to the action of heat, and the various attempts at the treatment of local tuberculosis by heat have been so encouraging as to merit continuance.

Bicycle and other accidents.

3. In the treatment of sprains, accidents which have become frequent since the wide introduction of the bicycle.

Chronic ulcers.

4. In the treatment of certain atonic ulcers, where healing is hindered and retarded by malnutrition of the adjacent and surrounding tissues. The use of heat in the treatment of such cases has been tried elsewhere, and certain patients suffering from varicose ulcers have been benefited by the application of compresses soaked in water almost at the boiling-point We have read of some English cases in which chronic ulcers have been treated by the Tallerman method with a success which we, unhappily, have been unable to attain.*

Pelvic regions.

5. Thanks to Mr. Tallerman's apparatus for the pelvic region, lumbago can be treated as well as the diseases of the hip and abdomen (coxalgia, sacrocoxalgia), (Brodie's disease).

Hysterical motor and sensory.

6. I mention here hysterical coxalgia, for I believe that these hot-air baths can with benefit be added to the means which have been used to treat the painful or paralytic symptoms, both motor and sensory, of hysteria.

7. In conclusion, there is a disease with which a few experiments will prove interesting. I refer to soft chancre.

If it is true that the Ducrey-Unna bacillus succumbs to high temperatures, it is possible that we may obtain results with the Tallerman apparatus which have not yet been obtained in the treatment of soft chancre by prolonged hot baths, hot applications, etc.

Although the list of diseases treated by this local application of superheated dry air appears somewhat long and diverse, we must say that there is no intention to claim that this treatment will cure all diseases. Such a pretension has never been made.

---

* Further experience in the use of this method has been followed by successful results; see photographs taken in the Liennec Hospital of later cases, "The Tallerman Treatment," published by Baillière. Tindall, & Cox, London.

All that we can say is, that it seems to be destined to render great service in the hands of the physician and surgeon, for, on the one hand, it has taken effect on diseases reputed to be incurable, such as chronic. deformatory rheumatism, against which the physician up till now confessed himself to be completely powerless; and, on the other hand, it has acted more quickly and with better results than all other therapeutic methods in diseases often intractable, such as sciatica, gout and gonorrhœal rheumatism.

The Tallerman treatment will render great assistance to physician and surgeon, and take effect in incurable diseases in which the physicians have to confess they are completely powerless

NORTH-WEST LONDON HOSPITAL.

Report and Notes on Cases treated by the Tallerman method Sprain. Acute attack of Gout in big toe, following on Sprain of Ankle. Chronic Rheumatoid Arthritis. Old Tuberculous Knee-joint. Chronic Ulcer: and Remarks by Editor of *Lancet,* J. F. SARGEANT, M.R.C.S., L.R.C.P.

(See Lancet, 12th Jan. 1895.)

THE authorities of the North-West London Hospital having at the suggestion of Mr. Mayo Collier kindly consented to submit a series of cases to treatment by the "local and medical dry-air bath," the proprietors of that apparatus courteously placed one of their cylinders at the disposal of the medical staff for that purpose, and I give the notes and results obtained during the past two months. It might not be out of place to refer here to the fact that this apparatus was introduced to the medical profession by Mr. Willett at St. Bartholomew's Hospital, in the wards of which institution a variety of cases were treated for two months, after which Mr. Willett detailed the results obtained in a clinical lecture delivered on May 23rd last. Acting upon the suggestions thrown out by the lecturer on that occasion several medical cases were treated. Under treatment patients experience a sense of comfort, probably due to the high temperature exercising an anodyne influence, which relieves the pain, or more often removes it entirely; even when adhesions have been broken down the pain is much modified if the joint is immediately subjected to treatment. Some of the cases can hardly fail to be of interest, being of that chronic character for which so little can be done by ordinary medical treatment; all of them were selected for their severity in order to test to the utmost the value of the apparatus.; the majority were cured and the remainder exhibited such marked improvement that it is only fair to state that there was not a single failure. Mr. Lewis A. Tallerman gave his personal attention during the treatment and Mr. Mayo Collier supervised the selection of cases.

Cases selected for their severit Not a single failure.

CASE I.—*Sprain.*—Child 11 years. Right ankle swollen and painful; could not put foot to the ground; effusion into tendon sheaths and joint itself. August 13th: treated forty minutes; pain and effusion considerably lessened; could put foot to ground. Treated 14th and 16th: patient walked well without pain. Discharged cured. Severe sprain cured in three operations given in four days.

CASE II.—*Sprain and Gout.*—Man, 57. Sprained right ankle. Next day gout attacked great toe of the injured foot. Was in great pain and had not slept for two nights. First operation of forty minutes. Pain modified. This patient was cured of a bad attack of gout with severe sprain in five operations given in twelve days.

CASE III.—*Chronic Rheumatoid Arthritis.*—Woman, 64. Knees, shoulders, wrists and fingers affected; could not feed or dress herself. Under medical treatment for four years, the disease making rapid strides notwithstanding. August 10th. First operation: Marked improvement. August 24th. After 9th operation: Patient reported she had returned to work and could walk up and downstairs without pain.

This case of chronic rheumatoid arthritis had been under continual medical supervision from its onset, and the usual remedies were applied. Its history shows that, so far from yielding to treatment, the disease made rapid strides, so that whereas eighteen months ago the patient was able with some

effort to follow her occupation of dressmaker, twelve months later she was incapacitated from even feeding or dressing herself. The improvement wrought by the hot dry-air cylinder was as immediate as it was remarkable; the progress of the disease was arrested, and a curative process was set up at

the first operation, which became more manifest with every succeeding one, proving beyond doubt the value of the treatment in cases of this nature.

CASE IV.—*Old Tuberculous Knee-joint.*—Patient 12 years. Joint stiff after being kept in plaster of Paris. Treated six times. Gained 45° to 50° in movement. Patient walked freely and well. No force used.

CASE V.—*Chronic Ulcer of Leg.*—Woman 45. Ulcer size of palm of the hand, edges unhealthy and thin, white slough. Loss of tissue 3/16th inch deep. Two operations: ulcer cleaner, edges healing fourteen days later, lost tissue was replaced and ulcer filled up. Margins healing.

CASE VI.—*Chronic Rheumatoid Arthritis.*—Man 71. Both elbows and all finger joints stiff and painful. Worked with greatest difficulty. Treated five times. The improvement as to pain and stiffness so marked, that he worked without inconvenience.

CASES VII. AND VIII.—*Two Cases of Sprained Ankles.*— The pain in both had disappeared after second treatment and joints were quite sound after fifth.

Two severe sprains, joint sound and discharged cured after five operations. Treatment removes at least in part a reproach from surgery. ED.—*Lancet*

Remarks.—It must be confessed that the results obtained by the usual treatment in cases of chronic stiffness of joints are far from satisfactory, so that one is inclined to welcome all the more cordially a recent therapeutic method which claims, and apparently with justice, to remove, at least in part, this reproach from surgery (ED. *Lancet*).

THE TREATMENT OF RHEUMATIC AFFECTIONS BY THE TALLERMAN METHOD.

(From the Medical Clinic of the Royal Victoria Hospital.)

By JAMES STEWART, M.D., Professor of Medicine and Clinical Medicine, McGill University; Physician to the Royal Victoria Hospital, Montreal. And W. G. REILLY, M.D., Senior Resident Physician, Royal Victoria Hospital.

IN December last, Mr. Lewis A. Tallerman, of London, gave two demonstrations of the method of using his hot-air apparatus at the Royal Victoria Hospital before a large number of the practitioners of Montreal. Since then the apparatus has been in constant use in the treatment of various forms of subacute and chronic rheumatic affections. The results on the whole have been very satisfactory. Relief to pain has usually followed, and in nearly all cases there is soon noticed not only an improvement in the local conditions, but also a marked change for the better in general nutrition. In the present preliminary communication an account is given of three cases treated to a conclusion. *In all it will be noticed that the results obtained are much more marked and satisfactory than by any other method at present known.*

Results more satisfactory than by any other method known.

Marked improvement in general nutrition.

CASE I.—Subacute Rheumatism of Four Months Duration— Multiple Arthritis, involving chiefly the Shoulder, Knee, and Vertebral Joints—Received Thirteen Hip Baths—Left the Hospital Greatly Relieved—The General Nutrition Much Improved.

CASE I.—A. T. L., æt. 42, was admitted to hospital on December 30th, 1896, complaining of rheumatism, he giving a history as follows : About the middle of September, 1896, he suffered from headache and feverishness. Towards the first of October pain and swelling set in in the left ankle, and still later in the right

Case of Rheumatism with great pain—emaciation, anæmia, unable to walk without crutches. Cured in thirteen operations.

ankle and in both knees, giving a clinical picture of acute rheumatism. For about a month he suffered acutely, then the pain disappeared to some extent, but the patient was unable to walk, while with the disappearance of the pain in the joints above mentioned, he complained of pain in the spine and pain and stiffness in the shoulder joints, so that motion was almost impossible.

Of his personal history it may be said that he has had repeated attacks of acute rheumatism, and in addition has had much mental worry consequent on business difficulties.

On examination after admission, the patient was seen to be a rather *emaciated, poorly-nourished man, with some anæmia,* unable to remain in any position for any length of time, and suffering great pain on movement. He was unable to walk without crutches, and then only with great difficulty. The right shoulder joint was very painful on manipulation, and movement in all directions much limited ; the same, but to a lesser extent, being the condition of the left shoulder joint. Motion of the vertebræ caused severe pain, chiefly in the lumbar regions. The knees were semi-flexed, and could be with great difficulty extended, but could be fully flexed. There was nothing abnormal about the ankle joints. In addition to the above there was a faint systolic murmur at the apex, and the pulmonary second sound was accentuated. The other viscera were normal.

The patient was treated by the Tallerman Apparatus, and had in all thirteen baths, with the result that at the time of discharge the pain and stiffness had been entirely removed from the knees, and the patient could walk without aid, but he still had slight pain in the back, and the right shoulder was a little stiff but not painful ; the left was free from both pain and stiffness. *The appetite and general health were much improved,* notwithstanding that the patient had a severe attack of tonsilitis while in hospital.

Gonorrhœa arthritis, pain, swelling, unable to stand or walk. Decided improvement after first treatment, walked after third. Gain in weight under treatment.

CASE II.—Gonorrhœal Arthritis of the Right Knee of Four Weeks Duration—Inability to Walk or Even Stand—After the Third Bath he was Able to Walk About the Ward Without Assistance—After the Twentieth Bath, he was Discharged Free from Pain and with Good Movement in the Joint.

CASE II.—J. L., æt. 22, was admitted to the hospital on December 18th, 1896, complaining of pain and swelling in the right knee, with inability to walk.

The history of the case is briefly as follows : In October, 1896, he contracted gonorrhœa, and up to the middle of November had a profuse urethral discharge. At this time the knee suddenly became much swollen and intensely painful ; the conditions lasting about three weeks, when it became less painful, but was still swollen and very stiff. There was no constitutional disturbance other than rapid loss of flesh. Of his personal history nothing was ascertained save that he had used alcohol and tobacco to excess.

On admission the patient was found to be unable to stand without support, and could not walk at all. The knee could be flexed and extended but a short distance, while there was uniform enlargement of the tissues about the joint, but no increase in the synovial fluid. No other joint showed abnormality. The urethral discharge contained gonococci.

The treatment, as in the previous case, was the use of the hot-air bath. After the first application a decided improvement was noticed, while after the third the patient was able to get about the ward without assistance. In all he had twenty applications, and on discharge, January 27th, 1897, there was free movement of the joint and practically no pain in walking. Although there still remained some enlargement and deformity about the joint, his *general health had much improved as evidenced by a gain in weight of* 14 *lbs. during the period of treatment.*

CASE III.—Repeated Attacks ot Subacute Rheumatism— Anæmia—Emaciation—Arthritis of the Right Knee and Shoulder—Marked Limitation of Movement in Several Joints —Air Treatment Quickly Followed by Relief to Pain— *Marked Improvement in General Nutrition.*

Rheumatism, anæmia, emaciation, painful joints and fluid in knee, progressive weakness and loss of flesh. Speedy relief, marked improvement in general nutrition, rapid gain of weight, patient discharged well.

CASE III. — J. H., æt. 26, was admitted to the hospital November 3rd, 1896, complaining of pain in the back and joints and difficulty in walking.

The onset ot the illness is stated to have been six years previous, at which time he suffered severe pain in the right hip joint on movement or pressure. Two months later he had acute inflammation in the right knee and subsequently the ankles and shoulders became involved successively.

After being in bed for nearly a year, the patient began to move about on crutches, and, still later, was able to get about when using a cane, but not otherwise. The condition remained stationary, apparently, for the next four years, then he began to suffer from pain in the back, with progressive weakness and loss of flesh. In December, 1895, he had acute inflammation in the right knee, and this has persisted to some extent up to the present time.

Save that he had a great deal of domestic worry, his personal history was negative.

On examination, he was found to be *emaciated and anæmic.* There was enlargement of the right knee joint, due to fluid, and the joint was hot and painful, and, in addition, there was some enlargement of the right shoulder joint. There was limitation of movement in the left hip, and in the left knee and right hip pseudo crepitus was easily demonstrated. The patient could not walk without a cane, and, when walking, there was limitation of movement about the pelvis, the trunk much bent forward, the feet widely apart, and he could not stoop to pick an object off the floor.

The examination of the viscera revealed nothing abnormal· This patient is still under treatment, but after the first application an improvement was noticed. He can now walk about without a cane, has become much more erect, and can pick an object off the floor readily. The effusion into the joints has disappeared and there is practically no pain either in the joints of the extremities or in the back. *His general health has much improved, there being a much better appetite and a rapid gain in weight.*

On February 20, patient left the hospital practically well.

*   *   *   *

Remarkable therapeutic effects obtained by the treatment hitherto regarded as impossible.

The therapeutic effects produced are relaxation of the part, copious and free perspiration over the whole of the body, enormously increased circulation and raising of the body temperature of from $1\frac{1}{2}$ to $4°$ F. This last effect, so contrary to the belief hitherto held that the body temperature could not be raised by a local application of heat, is remarkable, and, it is the belief of Tallerman, will before long be shown to have a very material and beneficial effect in the treatment of diseases other than in the classes of cases which until now have been subjected to it.

*   *   *   *

Safety of the treatment and benefit obtained when heart and kidneys are affected.

It has been proved that the treatment can be safely administered with benefit even where great debility, weak action or valvular disease of the heart or kidney disease are present. Rheumatic and other pains are relieved if not entirely removed shortly after the commencement of the first operation, and that the treatment itself is not only absolutely painless but so soothing as to frequently lead to patients falling asleep if permitted whilst under it, hence the sleeplessness caused by rheumatic pain is relieved and patients are able to rest at night.

SIXTY-FIFTH ANNUAL MEETING OF THE BRITISH MEDICAL ASSOCIATION.

Held in Montreal August 31st, September 1st, 2nd and 3rd.

Proceedings of Sections. Section of Medicine. STEPHEN MACKENZIE, M.D., President.

Discussion on the Relation of Rheumatoid Arthritis to Diseases of the Nervous System, Tuberculosis and Rheumatism.

BY JAMES STEWART, M.D.

Professor of Medicine and Clinical Medicine McGill University ; Physician to the Royal Victoria Hospital, Montreal.

British Medical Journal, 30th October, 1897.

IN my opinion the most valuable of all methods of treatment is the use of baths of superheated dry air, after the Tallerman method. It has been used in twenty cases of rheumatoid arthritis in the Royal Victoria Hospital during the past nine months with gratifying results. *The superiority of the Tallerman treatment over all other methods*

* * * *

In all we have treated twenty cases with the hot-air bath. In fourteen of the twenty cases pain in the affected joints was present and of a severe character. In the great majority of the cases the relief was marked even after the first bath, and after several baths the patient, except on movement, was practically free from pain. As a result of this relief, sleep, which usually before was greatly disturbed, was possible. In addition there was some apparent change for the better in nutrition. In spite of losing daily more than a pound in weight from the loss of fluid by perspiration, the patient usually steadily gains in weight. Gains of from three to four pounds weekly have been common. As regards the effect on the affected joint it is various, depending on the amount of effusion and the degree of ankylosis. *Twenty cases of rheumatoid arthritis treated with gratifying results at Royal Victoria Hospital* *Patients sleep better*

Generally a considerable increase in the mobility follows after the use of a few baths.

It cannot be expected that restitution can take place in advanced cases, but before much actual destruction takes place there is every reason to look for a decided check to the progressive character of the disease.

MEDICAL CONGRESS, BERLIN, CONGRESS, 1897.

Report from the Proceedings of the 15th Congress of Internal Medicine. Edited by Professor VON LEYDEN and Dr. EMIL PFEIFFER. Chronic Articular Rheumatism and its Treatment. By Dr. ADOLF. OIT, of Prag-Marienbad.

A SPECIAL apparatus designed for local hot air baths has recently been constructed in England by Mr. Tallerman, and has

already been used in various hospitals with the greatest success.
The temperature can be easily read off from a thermometer let
down into the chamber.   The apparatus is heated by gas
burners arranged underneath.   On introducing the limb, the
temperature is brought up to 150° F. and then raised to 220°
and even 300°.   The consequence is a profuse outbreak of
perspiration in the enclosed limbs, extending, however, also over
nearly the whole body.   Often after the first sitting the pains
are materially relieved, the joints more supple, the movements
freer, and that not only in the parts directly submitted to the
hot air but also in the rest of the body.

The improvement effected by the apparatus in the cases
described is in fact so astonishingly great, that we have hardly
ever seen it equalled by any other means.   Cases of very long
standing—thirty years in one instance—and crippled in an

Long lost use of limbs restored. extreme degree, were so far improved that the pains disappeared
and the advanced immovability of the joints yielded to such a
remarkable extent as to restore the long lost use of their limbs
to the sufferers.

MEDICAL SOCIETY OF BERLIN.

From the *Deutsche Medicinische Wochenschrift,*

March 17th, 1898.

Demonstration by Dr. MARTIN MENDELSSOHN of the method of
treating chronic articular rheumatism by superheated dry
air—(TALLERMAN METHOD).

GENTLEMEN: I beg to bring before you this method of treating
articular rheumatism, which is based upon the local action of
superheated dry air on the affected and deformed limb.   The
numerous and varied methods of treating chronic affections of
the joints, and particularly polyarthritis deformans, have, so far
as they consist of a local or general action on the surface of the
body, nothing specific about them, but in all of them the real
and efficient factor is the thermal effect.   And one may say that
those methods are the most efficient which enable this heat-
action to be applied in the most intense form and for the longest
time, while excluding to the utmost any unfavourable reaction,
especially upon the heart and nervous system.   We know that
the action of heat in this connection consists in the stimulation of
the skin and consequent relaxation of the subcutaneous vessels.
By this means the deeper lying vessels are contracted, and the
internal organs de-congestionized.   Moreover, by the extraordinarily powerful effect upon the movement of the venous
blood and the lymph, resorption is greatly promoted.   All the
usual methods of treating chronic articular rheumatism agree in
these points; the only question is how to apply the heat in the
highest possible degree without causing any local or constitutional injury.   This may occur on the one hand in the shape of

scorching, and even burning, which makes further treatment impossible; or on the other by disturbance of the whole organism as in steam baths, by increasing the frequency of the pulse and respiration, by engorging the cutaneous vessels and increasing the reaction of sweat to such a degree that persons who are not very strong cannot bear such demands, especially upon the circulating system, without ill effects.

The apparatus which I beg to show you here avoids those difficulties very ingeniously. Mr. Tallerman of London has conceived the idea of employing superheated dry air, locally applied. You see before you two apparatus, which serve this purpose, one designed for the upper extremities the other for the lower and pelvis. They consist of a copper cylinder or chamber in which the affected limb can be enclosed. At the beginning of the sitting, when the extremity is introduced into the apparatus, the temperature is 65° C.; it is then raised to 100° C. and even to 120° C., but, as I have repeatedly convinced myself, it can be brought up to 150° C. and higher without causing the patient any harm or even any unpleasant sensation. This method of applying heat has the very great advantage of permitting the use of a much higher temperature than can be approximately reached by any other. It is also continuous throughout and—a very important point as regards the effect—can be maintained for 40 and 50 minutes at a stretch. In addition the action is purely local, the patient breathes the cool air of the room and is not injuriously affected by the heat; especially a harmful reaction on the heart is avoided so that the treatment is applicable even when cardiac weakness is present. And finally—quite apart from the simplicity of the whole arrangement and its universal applicability—the effect is not confined to the part treated. The other joints though in a less degree become more moveable and less painful, obviously in consequence of the increased circulation and absorption, principally due to the free withdrawal of moisture through perspiration, which is continually maintained by the renewal of the air in the chamber. Frequently after a single good long sitting the joints are found to be relaxed and the movement freer; and in many instances the effect is perfectly astonishing, fixtures of the joints of long standing being so far improved as to restore the use of the limbs. Of course such a remarkable result, obtained from a single sitting, is not permanent unless the process is repeated. By systematic use of this dry heat, however, results may be obtained which can hardly be approached by other methods.

Editorial Note.—On Thursday, February 24th, 1898, Mr. L. A. Tallerman demonstrated his treatment on some patients in my wards before the members of the Congress. A patient with chronic arthritis, who had been unable for a year to flex the fingers on the palm nearer than 2-3 centimetres, was actually able to close the hand completely after 30 minutes treatment at 120° C.

Marginal notes:

The Tallerman Treatment explained.

Treatment applicable when cardiac weakness is present.

Astonishing effects on fixed joints, results obtained hardly approached by other methods.

Mr. Tallerman's demonstration before the Congress, remarkable result obtained in a case of chronic arthritis.

MEDICAL CONGRESS, WIESBADEN.

Proceedings of the Sixteenth Medical Congress held in Wiesbaden, April 13-16th, 1898. Edited by Professor VON LEYDEN and Dr. EMIL PFEIFFER.

THE THERAPEUTIC APPLICATION OF VERY HIGH TEMPERATURES. (TALLERMAN METHOD).

By Dr. MARTIN MENDELSSOHN (Berlin).

AT the Medical Congress held last year in Berlin, the principal discussion took place on Chronic Articular Rheumatism and its treatment. Among other special methods one by superheated air was mentioned as having lately been brought out in England. The few and exclusively foreign accounts of it which had appeared credited it with results not merely favourable but altogether extraordinary and frequently even astounding. Nevertheless, here in Germany, nothing was known of this method of treatment with dry air at extraordinarily high temperatures. When, therefore, Mr. Tallerman, the inventor of the method, offered to place his apparatus at my disposal to test its efficacy, I thought it worth while to subject the use of these high temperatures and their actions on the human organism to a thorough trial. For apart from the really enthusiastic testimony of English and French writers of high standing, the method seemed *a priori* to be necessarily advantageous and in the circumstances superior to others of a like character, provided only that the first premiss were fulfilled, namely, the practicability of subjecting patients without ill-effects to such temperatures, which are far in excess of those usually applied for therapeutic purposes. I have accordingly devoted myself, in the first instance, to settling this purely physiological question by actual trial, and beg leave to confine my remarks to-day to that preliminary point, reserving the discussion of my therapeutic experiences for the present. After an exhaustive trial, I can state with absolute certainty that temperatures of 120° C. to 140° C.—that is to say, temperatures which other methods cannot even approach—can be not only applied with the greatest ease by means of this apparatus, but also borne by the patients without any discomfort or ill-effects. This tolerance obviously rests on the special character of the apparatus, which distinguishes it from all others of a similar kind—namely, the property of keeping the air perfectly dry and therefore of permitting the maximum amount of evaporation

The physiology of the Treatment practically determined.

Temperatures used which other methods cannot even approach.

of perspired moisture. The physiological effects depend more upon this intense perspiration than upon the immediate action of the high temperature.

The possibility of keeping the air dry in the apparatus depends upon its peculiar yet extremely simple construction. The important point is the arrangement for ventilation by means of valves, which, while maintaining a constant temperature in the interior, permit the moisture-laden air continually to escape and be replaced by dry air. It is obvious that, in consequence of this process, the perspiration set up in the enclosed limbs by the extremely high temperature within the apparatus, must be developed in a quite exceptional degree—indeed to the utmost possible limits—and that in this way a greatly increased discharge of moisture is effected through the skin, and removed by the stream of air through the apparatus, as it is well known that the rate of evaporation depends not only upon temperature but very much on the atmospheric pressure, and particularly on the quantity of moisture present in the air. This profuse evaporation from the surface of the enclosed limb alone renders it possible for the living organism to bear such high temperatures, by producing a coolness which serves to lower the temperature of the surface to somewhere about normal.

In this connection I may mention the fact, which I have established by a great number of observations, that the body temperature only undergoes a slight rise in this local application of hot air, even when a temperature of 140° C. is applied for two hours and more. The pulse also shows only slight difference : from four to eight beats in the minute is the greatest increase in frequency I have observed, and that is generally accompanied by increased fulness and strength.

Without going into figures and tables on this occasion I am able to state with certainty, from the very large number of my observations in the most varied classes of patients, that the local application of excessively high temperatures is not only possible but can be carried on with perfect safety for a comparatively long time, if it is considered necessary, even for several hours continuously. And here we have once more an illustration of a truth which has not seldom been established in the history of therapeutics—namely, that the human organism possesses a capacity for enduring treatment, which medical men with their professional responsibility must shrink from attempting, and the first trial of which must therefore be undertaken, as in this case, by laymen with their happy freedom from such shackles. But supposing the venture succeeds ! And it is the fact that the organism is able to stand the action of these high temperatures. *That excessively high temperatures can be administered with perfect safety, established by observation.*

Now if it is heat that plays the principal part in all the physical methods of treatment of this kind, then this particular application of dry hot air can, *a priori*, claim superiority and *Testimony to the superiority of the Tallerman method over all similar ones.*

CASE 2.—Arthritis deformans, taken April 22nd, 1896, before treatment, patient aged 69: showing enlarged joints and fingers fixed in above position, cannot flex or extend nor use the hands for anything.

For Case notes and Report, see page 17, "The Tallerman Treatment" (Baillière, Tindall, & Cox, London).

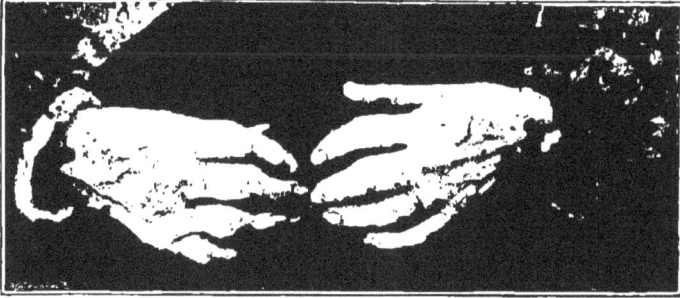

CASE 2.—Taken April 24th, 1896, after second operation : showing some separation of fingers, and furrows in the neighbouring fingers caused by the pressure of the enlarged joints.

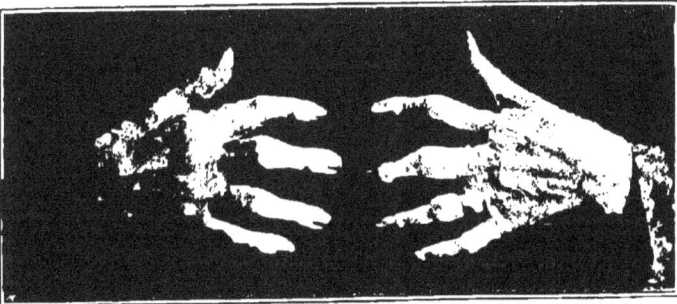

CASE 2.—Taken August 7th, 1896: showing result of treatment, fingers can be flexed or extended, movements quite free and comfortable.

For Case notes and Report, see page 17, "The Tallerman Treatment"
(Baillière, Tindall, & Cox, London).

D 2

precedence over all of them in so much as it permits the use of very much higher temperatures and for a very much longer time. No doubt other elements come into play in the various methods of treatment, but the thermal effect is certainly the most important, and it reaches its highest point in this superheated dry - air method which we have before us. Moreover, in the simplicity and cleanliness of its use, in its universal applicability and ready handling, in the exactness of its local adaptation, in the precision and ease with which it can be regulated, it possesses real and valuable advantages.

Enables the use of much higher temperatures over much longer time.

The mode of its action is of course similar to that of every local heat-application, though heightened in degree. Enlargement and relaxation of the superficial bloodvessels; their engorgement, with consequent withdrawal of blood from the deep-seated organs and contraction of the deeper vessels—these are effects which must increase the movement of the venous blood and of the lymph to an exceptional degree under the circumstances. This increased movement of the fluids in the body is further assisted by the profuse local perspiration, and absorption is thereby markedly promoted. The latter must be considered the chief result of these powerful thermal processes. The direct local over-heating, on the other hand, obviously falls into the background. There is therefore little probability that recent inflammation of the joints—as gonorrhœal rheumatism, for instance—can be cured by killing the *cocci* by means of high temperatures. If, nevertheless, in these acute cases a favourable result is obtained, as my incomplete observations shew to be probable, the mode of action is doubtless that with which we are so familiar in modern therapeutics. The cure is effected not by simply rendering inert the cause of the already established disease, but by the eliminations of the morbid products through the therapeutic measures applied.

Action of the treatment.

Inflammation of the joints cured by killing the cocci.

Cures are effected by eliminating the morbid products.

PROFESSOR BAUMLER. (FREIBURG).

Gentlemen,—I was very anxious for Mr. Tallerman to come to Berlin last year and demonstrate his apparatus then, when we were discussing chronic articular rheumatism and *arthritis deformans*. It was precisely cases of these extraordinarily intractable affections which had been originally treated in England with success, and indeed, according to what I have read, with great success. My attention was first drawn to his treatment by some cases published in the *Lancet,* and I took some trouble to induce him to come to Berlin ; but at that time he was unable to do so.

It is all the more gratifying that he has now come here and will presently give us a demonstration. Naturally we attach most importance to the evidence of our own eyesight, and wish to see in what manner the treatment is applied, and how a limb enclosed in the apparatus reacts to these excessively high temperatures. Dr. Mendelssohn has drawn attention to the thermal action of this local treatment, and has referred to the changes in the circulating system which must take place in a part of the body so treated. I believe that very great stress must be laid upon the latter in relation to chronic inflammations, for instance, of the joints; and I fancy that the present method will afford a parallel to that which Professor Bier of Kiel has recently applied with success to chronic joint affections.

CONGRESS INNERE MEDICIN, WIESBADEN.

Demonstration by Mr. TALLERMAN, assisted by Prof. BAUMLER (Freiberg), and Dr. M. MENDELSSOHN (Berlin).

Official Report :—*15th April,* 1898.

A DEMONSTRATION of the apparatus for hot-air treatment in affections of the joints, neuralgias and similar conditions, was given by Mr. Lewis A. Tallerman, of London, on April 15th. Geheim-Rath Bäumler and Dr. M. Mendelssohn gave the necessary explanations. Mr. Tallerman treated a patient, placed at his disposal by the directors of the hospital and suffering from a gouty-rheumatic affection of the wrist, for fifty minutes in the apparatus. The temperature of the air in the cylinder, in which the arm was placed, was raised to 116° C. Copious perspiration ensued, not only from the arm treated but from the whole body, which was wrapped in woollen coverings. The temperature under the tongue rose about 0·2° C. At the conclusion of the sitting, the patient, who had previously been unable on account of pain and stiffness to extend the half-flexed fore and middle fingers or to close the fingers on the palm, could do both without pain and also move the hand at the wrist joint. This result not only persisted on the following day but was even somewhat improved.

On April 15th another patient was similarly treated before a general gathering of members.

Mr. Tallerman's demonstration before the Congress of Innere Medicin, Wiesbaden. Splendid result obtained, improved still further by the following day.

FINAL MEETING OF THE CONGRESS.

Herr MORITZ SCHMIDT of Frankfort, in the Chair.

THE PRESIDENT: I have further to announce that Mr. Tallerman, who yesterday shewed us his hot-air apparatus, will present it to the Kaiser Wilhelm Infirmary for Invalid Soldiers, in recognition of the friendly reception accorded him. I am sure we thank Mr. Tallerman most heartily for having utilized the opportunity of our Congress to make such a valuable gift to our invalid soldiers.

Recognition of the friendly reception accorded by the Congress. Apparatus presented for the use of Invalid Soldiers.

THE GRAND DUCHY OF BADEN.

GOVERNMENT BATH ESTABLISHMENTS, BADEN-BADEN.

Report of the Adoption of the Tallerman Treatment.

"Badeblatt," 10*th September* 1898.

THE means of cure which, at the Bath Establishments of the Grand Duchy, have been placed within the reach of visitor to Baden-Baden, have been further enriched by a valuable addition.

We are to-day able to inform our readers of the introduction of a new cure, viz., "The Tallerman Treatment by Superheated Dry Air," respecting which the medical authorities report as follows :—

Tallerman Treat-
ment adopted on
the recommenda-
tion of the
German medical
authorities. The Ministry of the Interior of the Grand Duchy has, on the recommendation of the home medical authorities, introduced at the Government Bath Establishments, Baden-Baden, the treatment known as the *"The Tallerman Treatment by Superheated Dry Air,"* for chronic rheumatism, rheumatism of the joints, rheumatoid arthritis, gout, sciatica, and diseases of a like character, and during the past few days the necessary apparatus have been installed for general use.

The "Tallerman Treatment by Superheated Dry Air" was invented by Mr. Lewis A. Tallerman, of London, some time since, and that gentleman having completed his experiments and perfected his apparatus, introduced it to the public through the medium of the English medical profession. Communications which appeared in the English medical paper, the "Lancet," together with the book published this year upon the subject,* drew the attention of the German physicians to this new method, and as the communications were not only favourable, but also mentioned some very astonishing results that had been obtained, a great desire manifested itself to see and test both the treatment

The treatment
tested by the Ger-
man physician. and apparatus. This has now been done. Mr. Tallerman personally demonstrated both, first at the 16th Congress of Innere Medicin, held at Wiesbaden in April last, and again at the 22nd Meeting of the Neurologists and specialists in mental diseases, held at Baden-Baden in May.

The great value of the method lies in the very high temperature which can be locally applied ; this temperature, which the means hitherto at our disposal has not enabled us to approach, is effected by the Tallerman apparatus, not only without difficulty or drawback, but with actual comfort to the patient. A temperature of 140° C. can be administered without

---

* "THE TALLERMAN TREATMENT," edited by Dr. A. Shadwell, M.A., M.B. Oxon, M.R.C.P. Lond., published by Baillière, Tindall, & Cox, London.

producing any discomfort or feeling of intense heat; this is due to the moisture arising from perspiration being evaporated and the air in the apparatus being maintained dry. The perspiration of the limb under treatment is an extraordinary one, and by its evaporation creates an isolating zone round the limb, which so effectually protects the skin, that its temperature rises but very slightly above the normal, notwithstanding the high temperature in which it is immersed.

The physiological result lies not solely in the application of heat in itself, but also in the intense perspiration with all its beneficial effects on the whole organism and juices of the body.

It will be for medical men to indicate the cases which are suitable for treatment by the Tallerman method, in order that the hopes which rightly attach to its use may be realized.

To the Government of the Grand Duchy the warmest thanks are due, for its continued endeavours to bring within the reach of the sufferers who visit Baden-Baden new methods of cure such as this, which has been tested and approved by the medical authorities and has such a great future before it.

PHILADELPHIA COUNTY MEDICAL SOCIETY.

Meeting, November 11th, 1896.

Report of Meeting of Philadelphia County Medical Society. The President, Dr. J. C. WILSON, was in the chair.

Dr. HORATIO C. WOOD presented to the Society Mr. LEWIS A. TALLERMAN, who delivered a short address on the Tallerman Patent Superheated Dry-Air Treatment by the localized application of dry air heated to a high temperature.

Dr. H. C. WOOD said that Mr. TALLERMAN is the inventor of a new method of treating chronic rheumatism, rheumatoid arthritis, sprains, both acute and chronic, and a large number of general inflammations and rheumatic affections. It is proposed to put the apparatus to the test on two patients whose treatment will require from forty to sixty minutes, so that the results of the treatment can be seen.

Dr. FREDERICK A. PACKARD demonstrated a case of satur- *Lead poisoning,* nine gout that had been extremely obstinate to all forms of *obstinate case.* treatment, in so far as his articular symptoms were concerned, and had at the time suffered from severe pain in one of the big toes, which was constant in spite of all that could be done. On learning of the Tallerman method of treatment, Dr. PACKARD thought that the case would be a good one for demonstration, because the man was suffering pain in the big toe so that it could not be touched or flexed, and also because he had tender nodes on the hands, so that the effects on peripheral processes far from the seat of treatment could be seen.

Dr. H. C. WOOD presented a case of lumbago, in which it was supposed the apparatus would afford marked relief. After the demonstration Dr. WOOD pointed out that the man with the lumbago, who had moaned and groaned when touched, and had been in the hospital a week without much benefit, was, after the treatment by Mr. TALLERMAN, able to get up.

The case of saturnine gout had been under competent medical care for some weeks, without much gain. The man had not moved his toes for three months, but after the treatment he was able to move the toes freely. In fifteen minutes after the treatment was begun he could move his toe.

Most wonderful result.

Dr. H. C. WOOD said :—" It is the most wonderful result I ever have seen. This is a case of saturnine gout which is almost incurable."

MEETING OF PHILADELPHIA COUNTY MEDICAL SOCIETY.

Confirmation of the successful Demonstration at the above Meeting.

1,925 Chestnut Street, Philadelphia, Pa.

November 16th, 1896.

Professor Horatio C. Wood orders an apparatus and adopts the treatment.

My Dear Mr. TALLERMAN,—

It gives me great pleasure to state that, after trial of the Tallerman-Sheffield apparatus and perusal of the published literature of its use, it seems to me that this application of supra-heated dry air to joints bids fair to be an important addition to our therapeutic measures, and that your apparatus has worked satisfactorily.

Please let me have an apparatus (hospital model).

Truly yours,

Lewis Tallerman, Esq.   H. C. WOOD.

1,437 Walnut Street, Philadelphia, Pa.

November 20th, 1896.

LEWIS A. TALLERMAN, ESQ.

Improvement maintained.

Dear Sir,—I was much impressed with the immediate results of the treatment by your apparatus in two cases treated at the meeting of the Philadelphia County Medical Society on the 11th inst., and have since learned that the improvement was permanent

Very truly yours,

J. C. WILSON,

30th November, 1896.   Chairman of the Meeting.

110, South Eighteenth Street, Philadelphia, Pa.

November 28th, 1896.

To Lewis A. Tallerman, Esq.

Dear Sir,—In accordance with my promise to you, I write to report the condition of the patients treated in my wards at the Philadelphia Hospital by the use of your apparatus. Latham (the case with saturnine gout) continues in a comfortable condition and has had no return of his articular trouble. The case with chronic fibrous rheumatism of the knee is in about the same condition as when you left, being able to walk comfortably and to flex the leg on the thigh and the thigh on the abdomen far better than before the use of the apparatus. The large fat man with dyarthritis of the hip and sciatica still stays well and has not resumed the use of his cane since. The case of chronic rheumatism in Ward II. who had the stiff shoulder has retained all of the motion acquired by the application of the heat.

The other cases are in the wards of other visiting physicians and I cannot inform you of their condition. Dr. Spear could probably let you know how they progressed after your departure.

Very truly yours,

FREDERICK A. PACKARD.

SOCIETY PROCEEDINGS. PHILADELPHIA COUNTY MEDICAL SOCIETY.

Stated Meeting, November 11th, 1896.

Medical News, New York City, January 16th, 1897.

The President, Dr. J. C. Wilson, in the Chair.

Dr. H. C. Wood introduced Mr. Lewis A. Tallerman of London, who demonstrated the Tallerman patent local and medical dry-air bath, and showed the method of its application in the cases of two patients from the Philadelphia Hospital, one suffering from saturnine gout and the other from lumbago. Both were very materially helped, the patient with lumbago, after having suffered so severely as to require aid to get about, and then only with great pain, being able subsequently to the treatment to walk with ease and comfort.

*   *   *   *

The patient with saturnine gout was able to move joints that had not been moved for a long period of time, and not only those exposed directly to the influence of the apparatus, but also those at a distance and not treated directly.

THE LOCAL HOT-AIR TREATMENT.

Philadelphia Medical World, January, 1897.

A case of lumbago, and five cases chronic fibrous rheumatism greatly benefited by a single treatment.

THE results of the hot-air treatment by the Tallerman apparatus in this city have thus far been as follows: The case of lumbago, not chronic, treated before the Philadelphia Medical Society, was cured by that one treatment. The other case treated before the society, gouty feet, received one additional treatment and has been greatly benefited. One case of gonorrhæal rheumatism received two treatments without benefit. Five cases of chronic fibrous rheumatism received one treatment each and all were greatly benefited. The benefit received by this treatment seems to be permanent.

NORTH-WEST LONDON CLINICAL SOCIETY.

. Demonstration of Cases, Clinical Society, November 17th, 1897.
Dr. MILSON in the Chair.

A CASE OF ARTHRITIS ASSOCIATED WITH WARTS.

Clinical Journal, January 5th, 1898.

A cripple of seventeen years able to stand and walk without crutches after treatment.

Dr. KNOWSLEY SIBLEY showed a woman 56 years of age who for the last seventeen years had been a great cripple from rheumatism, but under the Tallerman treatment of hot dry-air baths she was able to walk with her head erect without either crutches or sticks. She presented an interesting condition of multiple warts, which had on more than one occasion more or less spontaneously disappeared.

BRISTOL MEDICO-CHIRURGICAL SOCIETY.

Annual Meeting, October 13th, 1898.
Bristol Medico-Chirurgical Journal, December, 1897.

Rheumatoid arthritis, use of limbs restored.

Dr. E. C. WILLIAMS showed two cases, fairly typical of the class of case which would be expected to derive benefit from the Tallerman treatment. (1) A boy, aged 8 years, was admitted into the Children's Hospital with a stiff left knee-joint and inability to flex the knee and walk. After six baths it was possible to slightly flex the knee, and now, after twenty-five baths, he is able to walk. (2) A girl, aged 17 years, with rheumatoid arthritis, with considerable effusion into the finger-joints ; the right elbow was flexed, and it was impossible for her to dress herself. After six baths the effusion was gradually absorbed.

Now, after thirteen baths, the patient is able to dress herself. The bath is generally given from forty-five minutes to one hour, and the temperature of the bath varies from 260° to 300° Fah. The body temperature of the patient is raised 1° or 2° Fah., and there is profuse sweating.

MEDICAL TIMES, May 22nd, 1897.

The Tallerman Hot-Air Treatment in a case of Chronic Rheumatism with Anchylosis of the Right Knee and both Elbow Joints. By W. KNOWSLEY SIBLEY, M.A., M.D., B.C.Camb., M.R.C.P.Lond.; Senior Physician to Out-patients, North-West London Hospital.

THE following case illustrates very well the striking effect of the treatment in a severe and practically incurable condition of deformity, produced by long-standing rheumatism :— *Striking result obtained in a practically incurable case*

The patient was a girl, aged twenty-six, single. Her mother suffered for fifteen years from rheumatism, and her father died, aged fifty-six, from kidney and heart troubles ; her maternal great-grandfather also suffered from rheumatism.

History.—Always well until about four years ago. She suffered from anæmia as a girl. The rheumatic affection commenced gradually in the right little finger, then left big toe, right knee, then the other fingers, and then hands and elbows in succession.

In August, 1894, she went to Bath and took the baths at the Mineral Water Hospital, and was there seven months —she had baths and salicylate treatment. She became worse, the right knee became contracted and fixed, so the baths were stopped ; she had a series of colds, and the rheumatism was worse. She returned to her home in Wales in April, 1895, then in many respects better, but shortly afterwards she again relapsed. *Seven months of Bath waters and drug treatment leave the patient worse.*

By November, 1895, she had become much worse, and was unable to open her mouth on account of rheumatism in the jaw.

In January, 1896, she went to Brecon ; while there was given colchicum, salicylates, iodide of potassium and tonics.

In September, 1896, she was sent to the Tallerman Treatment Institute, 50, Welbeck Street, London, for treatment.

State on admission, September 30th, 1896 :—Patient had used crutches for two years; she was unable to walk up and down stairs, to wash her neck, to put her jacket on or off, or to do her hair; she fed herself with a large spoon with difficulty as she could not get her hands to her mouth ; she could only see the back of her hands, and was quite unable to rotate the elbows. Pulse, 72, regular—no cardiac bruit.

Hands.—The middle phalangeal joints of both were considerably thickened, the right little finger was deformed at the terminal phalanx, wrists were thickened and the hands deflected outwards.

Elbows.—Both were almost fixed at right angles and completely pronated ; there was little or no movement of either flexion, extension or rotation.

Shoulders.—Considerable grating in both ; movements fairly free.

Right Knee.—Anchylosed nearly at a right angle; absolutely no movement ; muscles of this leg and also of the thigh much wasted ; the tip of the toes could just be placed on the ground ; the limb was powerless, patient could not even raise it off the bed without assistance, and there was considerable thickening around the knee joint, but there was little or no effusion. The tendons at the back of knee were very rigid, tense, and fixed. Patient was unable to place this foot on the ground without the guttapercha splint round the knee joint, and even then was unable to bear any weight on this limb. The knee was very painful.

Left Foot.—Œdema on dorsum.   Pain on movement.

The first hot-air bath was given on October 1st, the right arm being placed in the cylinder.

On October 3rd, after the second bath, it was possible to rotate the left elbow, so as to supinate the palm of the hand. After the third bath the patient was able to see the palm of her hand, which she had not been able to do for two years, and also to touch her forehead with it.

On October 7th, after the sixth operation, patient was able to do her front hair, and there was now some more movement of flexion and extension in the elbows.

On October 12th, after the tenth bath, the patient was able to walk a few steps without her crutches ; there was distinctly some movement in the right knee joint, and she was able to take a few steps upstairs.

On October 13th, the right leg was placed in the cylinder, and after treatment there was increased movement in the knee joint.

On October 14th, the patient walked out of doors for an hour, and on returning walked upstairs with the aid of one crutch.

By October 21st, patient had had sixteen operations, extending over twenty-one days ; the movements of the elbows and wrists were now sufficient to permit her washing and dressing herself, including doing her hair ; she was also able to feed herself with ease. All through the treatment she had been practically free from pain, even in the knee joint after one had forcibly broken down some adhesions ; these active movements had not been accompanied by any effusion into the joint. Her general health had also greatly improved. She has been taking a teaspoonful of the syrup of the iodide of iron three times a day and some natural saline water as a mild aperient, when necessary.

On October 22nd she had the eighteenth operation, and was shown before the North-West London Clinical Society ; she could then put the right foot more firmly on the ground. *Shown before the North-West London Clinical Society*

By November 4th patient could walk round the room without her crutches, with the help of a stick.

On November 6th, after the thirtieth bath, she was shown before the Harveian Society of London, and the movement, which was now to be seen in the knee joint, was demonstrated (*Lancet*, Nov. 21st, 1896). *Shown before the Harveian Society of London.*

On November 10th, it being thought desirable to hasten the increase of movement in the right elbow, patient was given a little gas and oxygen, and the elbow forcibly flexed and extended, and the same, to a less extent, was done to the right knee. As soon as she came round the right arm was put into the apparatus. She had little or no pain afterwards, and passed a very good night ; the next day there was no effusion into either of the joints which had been moved, and no pain about them. The increased movement was with difficulty maintained on account of the wasting of the muscles, especially of the biceps, through disuse. *Value of treatment where adhesions are broken forcibly; relief of pain.*

On November 20th patient's condition had steadily improved. There had been no rise of temperature or effusion into the joints since the movement under the anæsthetic. She left the home for a fortnight's change, and returned on December 14th, when she was able to walk up and downstairs without her crutches or stick.

On January 4th, 1897, after the fifty-fourth operation, she walked out without her crutches, only using a stick.

On January 15th patient left for a convalescent home in the Isle of Wight, where she continued to make steady, uninterrupted progress. She remained till the middle of April, by which time she could walk about without a stick for four or five hours a day, and she had had no return of rheumatism in any form.

The patient was also able to dress, undress, and in fact do everything for herself, and no longer considered herself a cripple or invalid.

CLINICAL JOURNAL, 31st July, 1895.

THE TALLERMAN PATENT DRY BATH.—The amount of benefit which has been derived from this treatment since it was first introduced to the reading medical public, in a lecture delivered at St. Bartholomew's Hospital, by Mr. ALFRED *Surprise how little the treatment is known.*

CASE 3.—Arthritis deformans, taken June 19th, 1896, before treatment: showing
(i.) highest point hand and arm could be raised; (ii.) inability to flex the
fingers; (iii.) expression of pain.

For Case notes and Report, see page 20, "The Tallerman Treatment"
(Baillière, Tindall, & Cox, London).

CASE 3.—Arthritis deformans, taken June 19th, 1896, after the first treatment, showing full extension of arm and forearm, flexion of hand, absence of pain.

For Case notes and Report, see page 20, " The Tallerman Treatment " (Baillière, Tindall, & Cox, London).

WILLETT, on May 23rd, 1894, has certainly been surprising; but not, perhaps, so surprising as the fact that it has not become even more widely known. . . . . The utility of this treatment for stiff joints, or for use after adhesions have been broken down, is enormous, the pain being greatly modified, if not entirely dispelled.

ROYAL PORTSMOUTH, PORTSEA AND GOSPORT HOSPITAL.

Report and Case Notes:—" Notes of Cases treated by the " Tallerman Dry-Air Bath at the Royal Portsmouth " Hospital.

" AT the invitation of Mr. D. WARD COUSINS, F.R.C.S. *(President " Council British Medical Association)*, senior surgeon to the " hospital, and through the courtesy of Mr. Tallerman, one of " the above cylinders for the administration of dry-air baths at " high temperatures has been placed at the disposal of this " hospital for the past month; and the undermentioned cases " detail the results obtained by the use of his apparatus. The " notes are furnished by Mr. T. H. BISHOP, M.B. & C.M. " Edin., House Surgeon, and Mr. H. W. MORLEY, M.R.C.S., " L.R.C.P., Assistant House Surgeon to the Institution.

" Royal Portsmouth Hospital.

" *November 5th,* 1894."

Stiff Joints after Injury.—S. C. Age 50. Married. Fell down stairs six weeks ago, and sustained severe sprain of wrist. Wore a splint for six weeks, and then consulted a doctor, who advised her to undergo treatment in hot-air cylinder. Came to hospital on October 8th.

On examination the wrist and fingers were found to be extended perfectly stiff, and attempts at passive motion caused great pain. Patient herself could not move any of the figers, but could touch base of index finger with the thumb. She was reluctant at having the adhesions first broken down under an anæsthetic, so that it was determined to try the effects of the air cylinder without.

First operation, October 9th.—Hand and arm placed in cylinder for forty minutes at a temperature of 240° F. After fifteen or twenty minutes she said she could move her fore finger slightly. On withdrawing the hand from the cylinder it was found that the fingers could be moved slightly without much pain, and she could herself touch the tip of her index finger with her thumb. Although before the treatment she complained of great pain on attempting to flex the wrist, after the bath it was partially flexed, and several adhesions broke with an audible snap, the patient experiencing but little pain.

Second operation, October 11th.—Forty minutes at 290° F. Movement in fingers and wrist still greatly increased. Patient could touch the tips of all her fingers with her thumb. Some more adhesions in wrist were broken without much pain.

Third operation, October 13th.—Thirty minutes, 280° F. Movement in fingers and wrist markedly increased. She can now flex all her fingers into the palm, and has a good range of motion in wrist joint.

Saw patient again on October 28th. She can now make a fist and move her wrist joint freely. Is greatly pleased at the result of the treatment, and has resumed her household duties. Pain on movement is but trifling.

Chronic Rheumatism, 18 *months.*—I. C. Aged 24. Carpenter. Invalided from Royal Engineers for chronic rheumatism. He has been unable to work for some months on account of pain in right wrist.

Right wrist: considerable thickening around wrist joint. Range of motion limited, and grasp very feeble. Attempts to move joint caused considerable pain.

First operation, October 18th.—Forty minutes, 240° F. Hand and arm placed in cylinder. After twenty minutes patient stated that he was quite free from pain.

Chronic rheumatism eighteen months—Pain relieved in twenty minutes

Second operation, October 20th.—Patient states that since the last operation he has suffered much less pain. The grasp is stronger. Range of motion increased, and the pain accompanying it is less. Placed hand in cylinder for forty minutes at 240° F. Thickening round joint decreasing.

Third operation, October 23rd.—The thickening round the joint has much decreased. Can now grasp firmly with the hand, and is almost free from pain. Hand placed in cylinder for thirty minutes, temperature 260° F.

Fourth operation, October 25th.—Temperature 260° F., forty minutes. Thickening almost gone. Pain very slight. Patient was unfortunately not able to attend any more. The improvement, however, was most marked, the thickening being all absorbed, the range of movement and strength of grasp being considerably increased, and when seen on November 2nd, he could flex and extend the wrist joint almost to their full extent without any pain, whereas, before the treatment, the movement was limited in extent, and accompanied with severe pain.

Fourth and last operation 25th October, most marked improvement; seen on 2nd November improvement had increased.

E

Stiffness after fracture. October 9th, third and last operation; return to normal, no pain. Nov. 2nd last seen, hand perfectly well.

Stiffness after Fracture.—Harry West, aged 12 years. Stiff fingers after fracture of forearm three months ago.

First operation, October 3rd.—Forty minutes at 265°F. Considerable improvement, fingers much less stiff.

Second operation, October 4th.—Thirty minutes at 240° F. The stiffness of the fingers has greatly improved, can fully flex the first and fully extend all the fingers with the exception of the little and ring, which cannot be quite straightened.

Third operation, October 9th.—Thirty-five minutes at 230° F. All the fingers can be fully flexed and extended without pain. When last seen (November 2nd) his hand was perfectly well.

Stiff Fingers after Collis Fracture.—Ellen Bundy. Aged 31. Nine weeks ago. All fingers are semiflexed. Unable to straighten them herself, and attempts to forcibly extend them caused severe pain.

Stiff fingers after Collis Fracture; mobility restored in four operations

First operation, October 4th.—Forty minutes. Temperature 250° F. After the bath the fingers could be further extended with less pain. Straight splint applied.

Operation repeated on October 9th, 11th and 16th.—On each occasion for thirty minutes, at temperatures of from 240° to 265° F. At the end of the treatment she could freely extend the fingers without pain, and she now resumed her occupation, and is quite well.

Rheumatoid Arthritis.—Miss M. Suffered with rheumatoid arthritis for some years, being a complete cripple. Knees and elbows stiff, and considerable deformity of the hands, especially the left, which is distorted and useless, the fingers being all flexed and stiff, and any attempt at movement causing severe pain.

Unfortunately the patient was unable to undergo full course of treatment, but the results of treatment, after two applications of the cylinder, were most encouraging. The fingers became more moveable and less painful. The little and ring fingers, which had been absolutely stiff and rigid before the treatment, became pliant and could with perseverance be almost straightened.

Chronic Traumatic Synovitis of Knee Joint.—J. L. Aged 49. Labourer. Patient had suffered from chronic synovitis of knee joint for over two years. There was a history of injury by fall on two separate occasions. Before the dry heated air treatment was commenced, other methods of treatment had been resorted to for several months without producing any satisfactory results. Before commencing treatment the knee was considerably swollen, was kept always in a position of considerable flexion, gave pain on movement and also when at rest, and had a very limited range of movement, not allowing of complete flexion or extension. The disease had therefore made considerable progress, and had affected the bones in addition to the synovial membrane of the knee joint. The hot-air bath was applied nineteen times. On the cessation of the treatment the circumference of the limb, from measurements taken above the patella and over the patella and below the patella had diminished to the extent of half-an-inch at each point of measurement. The range of movement also was slightly increased, though full flexion or extension still could not be effected. However, the pain experienced by the patient was undoubtedly considerably relieved.

This patient was under the care of Dr. D. WARD COUSINS, F.R.C.S., senior surgeon to the hospital, and the result of the treatment certainly was to render the condition of the joint more favourable for operative interference than it had been previously. *(marginal note: Condition of joint rendered more favourable for surgical interference.)*

NORTH-WEST LONDON HOSPITAL.

Notes and Report of the Tallerman Treatment in Acute and Chronic Gout, by W. KNOWSLEY SIBLEY, M.A., M.D., B.C. Camb., M.R.C.P. Lond., Senior Physician to Out-patients at the North-West London Hospital.

Lancet, July 10th, 1897.

I NOW wish to draw attention to the action and results of this treatment in a consecutive series of cases of gout, with the belief that the profession will welcome any advance in the present unsatisfactory method of dealing with a disease which directly or indirectly affects such numbers of individuals.*

CASE I.—*Acute Gout.*—The first case was a man, aged 65 years, who came under observation on June 3rd, 1896. The patient had suffered from acute attacks of gout on and off for many years, the attacks often lasting for several months, last attack lasted three months; present attack commenced few days before in left elbow, which became swollen—skin tense, red and painful, back of hand swollen, knuckles hardly visible, veins much engorged. First treatment, great improvement; treated our times in all; patient went into country for change. In April, 1897, patient had remained quite well and free from gout attacks since treatment. *(marginal note: Acute Gout treated four times in all.)*

CASE II.—*Acute, Sub-acute, and Chronic Gout associated with Eczema.*—This patient was a lady, aged 50 years, who commenced this treatment on June 25th, 1896. The patient when young had excellent health till the first attack of gout, twenty-four years ago, in the right great toe, when she was laid up for a fortnight. From this time she had recurrent attacks of acute *(marginal note: An old Chronic Gout case with Eczema, chalk stone, stiff joints, dislocated and painful, and discharging pus and chalky matter.)*

---

* For full Notes and Report upon these cases, see *Lancet,* 10th July, 1897. "The Tallerman Treatment," edited by Dr. Shadwell, M.A., M.B. Oxon., M.R.C.P. London. Baillière, Tindall, & Cox, King William Street, Strand.

E 2

gout every three or four months. For many years it was confined to the feet. It first appeared in the knees twelve years
previously and in the hands for the first time seven years
previously. The attacks had been much worse the last six years;
she was often laid up in bed for from six to ten weeks, and in
acute pain. Five years previously she had an attack of eczema
in the feet, and this had troubled her on and off ever since, never
having disappeared. Eight years ago the chalky gouty deposit
in the right great toe broke down and discharged for twelve
months, and this place had reopened on and off several times
since. In October, 1895, the deposit in the left index finger discharged for four months, and about the same time the left great
toe also discharged. From October, 1895, to June 25th, 1896,
the patient had been laid up in bed, being unable to move
because of pain, although taking large quantities of medicine.
The left index finger was much deformed, and chalky deposit
was seen beneath the thinned skin. The little finger of the
right hand was also much deformed ; there was chalk stone
deposit in the middle finger of the left hand. The left knee was
extremely painful and tender, movement was very limited on
account of pain, and there were some deposits to be seen and
felt on the anterior surface of the patella. Both the feet were
swollen and covered with an irritable form of eczema, which,
especially on the left, extended some way up the leg ; both great
toes were much deformed, tumid and inflamed, very red and
acutely painful ; they were both dislocated outwards and the left
was discharging thick pus and chalky matter and looked very
angry. There was considerable puffiness about the ankles and
pain on the slightest attempt at movement. There were deposits
in both ears. On June 25th the left leg was placed in the
cylinder. The pain was soon relieved by the heat, and when
taken out she was able to move the limb freely and could even
walk on it without pain ; the eczematous condition was also
better. The next day she reported that she had slept all night,
that the knee was almost well and the foot much less painful,
and that there had been no irritation from the eczema, although
for the first time for four months she had not used any ointment.
By January 21st the patient was well and she returned home.
The elbow had ceased to discharge and the movements of the
joints were unimpaired. She stated that the last time her
finger discharged it continued to do so for six months. In
April, 1897, the patient had continued quite free from gout since
January, and there had been no return of the eczema ; her
general health had also greatly improved.

Case of Gout
with Bronchitis
and Album-
inuria. After
first treatment
case improved
After third foot
quite well,
Bronchitis
improved, also
Albumen les-
sened.

CASE III.—*Sub-acute Gout, Chronic Bronchitis, and Albuminuria.*—This patient, who was first seen on September 21st, 1896,
had probably inherited gout from his mother's side of the family.
This patient, a painter by trade, had had severe colic fourteen
years ago, and was sixty-one years of age. The first attack of
gout occurred ten years ago in the right foot, when he was laid
up for a fortnight ; a second attack came on three years later,
and for the last three years he had had repeated attacks,

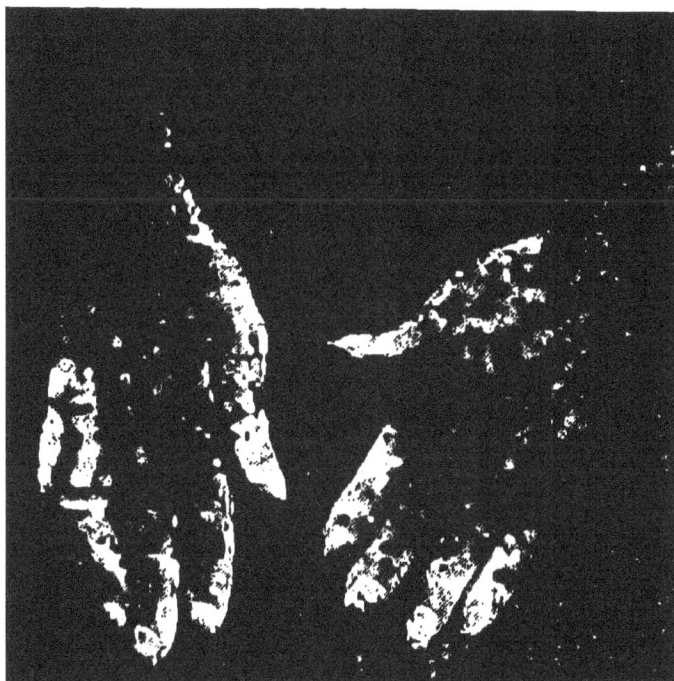

[Photographed in the " Laennec Hospital, Paris."]

CASE 6.—Before treatment : showing position of hands fixed against chest. Excessive pain, causing patient to cry out when he tried to move.

CASE 6·—After treatment (on eight occasions) : "the patient fed himself, which he "had been unable to do for a long time."

For Case notes and Report, see page 21, "The Tallerman Treatment" (Baillière, Tindall, & Cox, London).

especially in the ankles. When he came to the hospital the
right foot was swollen, very tender and painful, the skin being
red and much inflamed ; both legs were œdematous. Internal
remedies were prescribed. On October 12th the general condition was much the same. The cough and bronchitis continued
troublesome, with a good deal of expectoration in the early
morning. There were general bronchial rhonchi especially at
the right apex. The urine was clear, of specific gravity 1012,
and contained albumen. The second heart sound was accentuated. The condition of the foot was much about the same, the
patient being hardly able to get about, and then only with great
pain. On the 13th the first hot-air application was given, no
change being made in the medicines. The foot was much easier,
and he could move it more fully after treatment. By the 19th
he had had three baths. The foot was much better, and the
bronchitis was also much improved ; there appeared to be less
albumen than formerly. He had slept much better since the
baths were administered. On the 20th the patient stated that
his foot felt quite well ; he had no pain whatever. He took a
mixture of gentian and bicarbonate of soda all through the
treatment.

CASE IV.—*Acute Gout.*—This case of acute gout in a man,
affecting the left ankle, was greatly relieved by three baths.

CASE V.—*Acute Gout.*—A man, aged forty-three years, was
first seen on November 13th, 1896. His father died from rheumatism and Bright's disease. Gout occurred for the first time
ten years ago, and the attacks had been repeated at least once a
year ever since. This attack came on two days previously with
great pain, chiefly in the left foot. After third bath the foot felt
quite well, and the stiffness had disappeared and did not return.
On April 3rd, 1897, the patient wrote to say that he had not had
a return of gout in the feet and had continued at work ever since
the previous November.

CASE VI. *Acute Gout.*—A man, 24 years of age, came under
observation on December 7th, 1896. There was no family history
of gout or rheumatism. Fourteen weeks previously the patient
had suffered from inflammation of the eyes, which lasted three
weeks, followed by pain in the back, which kept him in bed for
three weeks ; afterwards the pain settled in the knees, the left
big toe, and then in the left hand and arm. Between December
7th and 21st he had four baths, and in addition a mixture
containing bicarbonate of soda and iodide of potassium. By
January 4th, 1897, he was quite free from gout.

CASE VII.—*Gout, Bronchitis, Heart Disease, Albuminuria.*—
This patient, who was a woman, aged sixty-five years, came
under observation on November 23rd, 1896. Her maternal grand-
mother, who suffered from gout, died when over eighty, and her
mother died at ninety-three years of age. She was quite well
till forty-three years of age, when she had the first attack of gout
in the left great toe. She had had frequent attacks since in the

feet, ankles, knees, elbows, and hands. For the last eighteen years she had also suffered much from bronchitis. On December 8th, she was admitted into the hospital. At this time the patient was very weak and suffered much from dyspnœa ; the pulse was rapid (120) and small ; the cardiac dulness was increased ; there was a soft systolic murmur at the apex and a much louder one at the base, which murmur was readily heard in the vessels of the neck ; the heart sounds were very feeble ; there were general bronchial rhonchi over both the lungs. The patient had had little or no sleep for weeks, and the breathing was rapid and shallow. On December 11th, after the first bath, the patient seemed to be a little easier. By the 14th the breathing was better after the bath, but she had not slept much By the 21st the right hand was not so painful and the cough was a little easier ; the patient still felt very weak, and her skin did not act well She gradually improved, and left the hospital on January 7th, 1897.

CASE VIII.—*Sub-acute Gout.*—The patient who was a cab driver, thirty-eight years of age, was first seen on January 4th, 1897. He was treated in the same way as the patient in Case VI. being given two baths, and by January 14th, he was free from pain.

Sub-acute Gout. Treated twice and discharged free from pain.

CASE IX.—*Acute Gout.*—The patient, a woman aged sixty-two years, came to the hospital on February 22nd, 1897. The present attack came on a week ago in the left shoulder, then the right leg, and three days later the right hand became involved. The whole hand was much swollen, very red, and the skin much inflamed and acutely painful ; there was some inflammation of the left ankle. She also suffered from chronic bronchitis. The first hot-air application was administered on the 23rd, with the result that the hand was less swollen and felt much easier after the bath. A second bath was given on the 25th. The fingers continued to be very stiff, but there had been much less pain in them. A third application was given on March 1st, when the patient said that she had slept much better and the pain had been much relieved. After this she considered herself cured and returned to her occupation on March 10th.

Acute Gout, patient 62. Treated three times. Patient considered herself cured and returned to work.

CASE X.—*Acute Gout.*—This patient was a salesman, who came to the hospital on March 21st, suffering from acute gout in both hands, which was quite relieved by three baths and an alkaline mixture.

Acute Gout in two hands. Discharged cured after being treated three times.

It is important to note that nearly all of the above cases were treated as hospital out-patients, and that therefore it was impossible to regulate or in any way control their daily habits as to diet, etc. Most of the patients were also only treated during an attack of gout ; it was difficult to persuade them to undergo treatment by the hot air between the attacks, even with a fair expectation of keeping off a return of the disease.

The results were obtained under great disadvantage, the patients being out-patients and irregular in attendance and diet.

* * * *

My experience with these cases is that directly the pain subsides the patient returns to work, and neglects to take medicine or the ordinary diet precautions. The cases above described contain examples of gout in its various stages, from the most acute, through the sub-acute, to the very chronic, these latter, sometimes with acute exacerbations, occurring from time to time. Taking first the very acute attacks (cases 1, 4, 5, 6 and 9), in all of these the local intense pain and congestion were almost immediately relieved by the hot air ; very obvious difference in the part was seen after the application, and the relief from the intense agony of an acutely inflamed gouty joint was most marked. Case 2 very well illustrates the treatment of very chronic gout, with gouty or chalk stone deposits ; the deposit in one finger, for instance, being so bad that at one time the medical attendant had wished to amputate the digit, and yet after several applications of heat the deposit gradually disappeared, and so also the deposits about the heads of the larger bones became distinctly lessened.

(margin note: Remarkable and immediate relief from intense agony and congestion.)

(margin note: A finger saved from amputation.)

*   *   *   *

This treatment applied locally to the inflamed gouty part causes an increased circulation in the area, bringing a larger quantity of blood to the seat of the lesion and at the same time taking away a larger quantity of blood probably more or less saturated with the bodies causing the local deposits. These deleterious products being in the general circulation, so to speak, dissolved out of the seat of the inflammation should in the normal course of events largely find their way out of the body through the kidneys, hence all should be done during the course of treatment to assist these organs to act as freely as possible. Some interesting observations on this part of the subject have recently been made in the clinics of Professor Landouzy at the Laennec Hospital, and Dr. Déjerine at the Salpêtrière Hospital, in Paris, who are experimenting with the Tallerman treatment, the results of which were published by Dr. Chrétien. They found, for instance, that in one case of ordinary gout the daily elimination of uric acid, which after the fourth bath was 57 centigrammes, rose after the ninth bath to 89 centigrammes ; and in another case of arthritis the daily co-efficient of urea had changed from 20 grammes 97 centigrammes before treatment to 25 grammes 50 centigrammes after the treatment had been administered. They also found an increased excretion of all the salts, especially the chlorides. This increased excretion through the kidneys is probably the explanation of the great benefit of this hot-air treatment in cases of gout. The results in the French hospitals were obtained entirely by this external method and without the exhibition of drugs of any kind. I feel no doubt of the great benefit this treatment will show in cases of what are described as " constitutional goutiness," without necessarily any local objective lesion.

(margin note: Though mixtures of various kinds were given in some of these cases, the French Hospitals obtained their results from the treatment only, and without drugs of any kind.)

(margin note: Great benefit of this treatment in " Constitutional goutiness.")

CHARING CROSS HOSPITAL.

Demonstration of Cases in the Wards, Thursday, January 27th,
1898, by FREDERICK C. WALLIS, B.A., M.B., B.C.,
F.R.C.S.Eng., Assistant Surgeon and Lecturer on Minor
Surgery at the Hospital.—*Clinical Journal*, March 9th,
1898.

WE have here in operation the Tallerman hot-air apparatus, _{Stiff joints made supple in half-an-hour.}
which you have all heard about and probably have seen. It
has proved to be very good indeed for stiff joints and conditions
of that sort. You will notice a thermometer inside. A temperature of 260° F. can be borne by most patients. Stiff joints put
into this become, after about half an hour, very supple. This
patient's wrist was very stiff when she put it in, and you will
now notice that it is quite the reverse.

USE IN FLAT-FOOT.

"The Deformities of the Human Foot, with their Treatment,"
by W. J. WALSHAM, M.B., F.R.C.S., Senior Assistant-
Surgeon, Orthopedic Surgeon and Lecturer in Anatomy
at St. Bartholomew's Hospital, &c.

" ' *Treatment of the Third or Rigid Degree of Flat-Foot by the* _{Saves an anæsthetic. Bones replaced and arch restored without force}
" *Hot-Air Bath.*—Since writing the treatment of flat-foot, we
" have, instead of giving an anæsthetic, placed the rigid foot
" in the Tallerman hot-air bath, in order to produce relaxation
" of the adaptively-contracted muscles and ligaments. In
" several cases in which the foot was held rigidly adducted with
" the bones displaced in the way characteristic of the severe
" degrees, and in which, moreover, the bones could not be forced
" back by manipulation with the unaided hands, we found that,
" after the foot had been three-quarters of an hour in the bath
" at a temperature of about 300° F., it came out quite supple, so
" that without any force the bones could be replaced and the arch
" restored. The foot was then in some instances put in plaster
" of Paris in the adducted and plantar-flexed position for the
" usual time, and in other instances a boot or boot-iron was
" ordered and exercises begun. ' "

MEDICAL PRESS, March 1st, 1896.

The Therapeutic Value of the Localised Hot-Air Bath.

THERE can be little doubt that in this local application of heated _{Most valuable therapeutic agent.}
air at a temperature of from 240° to 260° F. we have a most
valuable therapeutic agent. The baths known by the name of

Tallerman, accomplish this application of superheated dry air perfectly. The reports already before the profession of the results of the use of these baths in such affections as rheumatoid arthritis, chronic and gonorrhœal rheumatism, various forms of synovitis, acute gout, sprains, and in some orthopædic affections, have been, in a fair proportion of cases, most satisfactory. These baths have been used for some time at St. Bartholomew's, at the North-West London, and at Charing Cross Hospitals, while Dr. Ward Cousins has given them a trial, with considerable success, at the Royal Portsmouth Hospital. Also, demonstrations have been given in different places throughout the United Kingdom, and most favourable effects have been recorded in private practice. We have recently had an opportunity of seeing some cases that were submitted to this treatment, and we were much struck with the very great improvement in the reduction of deformity, and in the mobility of joints which followed the application of the heat. The bath has not been tried as extensively as it might have been in some types of skin disease. The softening effect it has on the skin in chronic cases of eczema and psoriasis would certainly indicate it as a valuable adjunct to other treatment, both through its action on the blood vessels and lymphatics locally, and in assisting in the absorption of other remedies.

Great improvement in the reduction of deformity and in the mobility of joints.

Softening effect on the skin in chronic case of eczema and psoriasis.

THE PROVINCIAL MEDICAL JOURNAL,

March 1st, 1895.

A formidable sprain cured in one operation.

On September 28th, a "Conductor" in the employment of the Brighton Omnibus Company, fell from the top of his 'bus and injured his wrist. On the following day he presented himself for treatment. The wrist was red and very much swollen, severe pain was complained of up to the elbow, in consequence of which during the previous night he had "no sleep." No fracture nor dislocation could be detected. The hand and forearm were placed in the cylinder for forty minutes at a commencing temperature of 160° F., reaching gradually 240. During the operation the man volunteered the statement that his wrist and fingers were getting more movable and less painful. On being released from the cylinder he said that "three parts of the pain had gone." He was to return on the following day. He did not do so, but on being seen at work on October 2nd, was questioned, and he explained that he did not come, simply because his wrist was practically well on the day after the hot-air bath, and quite well now.

* * * *

This case presented the appearance of a really formidable sprain, its rapid cure was very striking, and the result of treatment both in point of quickness and completeness, seems in this instance at least, far superior to that of any method with which the present writer is acquainted. Had there been fracture or dislocation as well as the symptoms named, it seems not improbable that an application of this kind would have proved a useful preliminary to reduction and fixation.

PHARMACEUTICAL JOURNAL, January 18th, 1896.

THE method has been tried in many of the large London hospitals, and the evidence of unimpeachable witnesses proves that for suitable cases it works successfully.

JOURNAL OF STATE MEDICINE, March, 1898.

THE Tallerman system of treatment of rheumatism, gout, rheumatic arthritis, sciatica, &c., by superheated dry air has been tried in various countries during the last four or five years, and numerous papers have been published by eminent members of the medical profession tending to prove the marked superiority of this treatment over any other used in various intractable forms of rheumatism. *Superior to any other treatment.*

*   *   *   *

As the diseases which appear to be most benefited by superheated dry air are exactly those which hitherto have almost defied all other methods of treatment, and as it has now been proved beyond all question that Tallerman's system is capable of affording marked relief in these saddest of all cases, it is obvious that there is not a general practitioner in the country who can afford to remain ignorant of its mode of application. *Ought to be universally known.* *Not a general practitioner can afford to remain ignorant of this system.*

THE HOSPITAL, February 16th, 1895.

ORTHOPÆDIC SURGERY.—A remarkable point about the treatment was its anodyne influence, movements that excited pain before can be performed after the limb is placed in the bath without pain.

THE HOSPITAL, August 1st, 1896.

Success of this new method warrants more detailed notice. IN January of this year there appeared among our book notices a review of a pamphlet dealing with the "Tallerman method of localised hot-air treatment." Since that date the experiences of other observers have been placed on record, and the success which has attended the application of this new method warrants more detailed notice.

*   *   *   *

Benefits derived, results from improved circulation. Rise in temperature an interesting phenomenon. The benefits resulting from this treatment are largely due to the vascular dilatation and improved circulation, not only in the part immediately exposed to the high temperature, but generally throughout the system. Counter-irritation and reflex influences may take some share in the matter. The rise in temperature is an interesting phenomenon, and at first appears in contradiction to physiological teaching. Although not explained by any of the observers who have used this method, it is probably due to imperfectly co-ordinated diaphoresis in parts of the body other than that exposed to the direct heat of the bath, that is to say the blood as it flows through the heated area is not completely cooled down before it passes on to other areas, where the conditions are different and where reflex superficial vascular dilatation and consequent diaphoresis are not correspondingly established. The improvement in condition is by no means confined to the part locally treated ; for instance, in rheumatoid arthritis the pain is alleviated, and the range of movement increased in distal joints also, and this not *Published cases show remarkable results, especially in cases of acute gout and rheumatoid arthritis.* only during but after the treatment. Many of the published cases show very remarkable results, and this is especially so in cases of acute gout and rheumatoid arthritis.

BRITISH MEDICAL JOURNAL, 16th June, 1894.

Anodyne effect specially striking in painful rheumatic joints. THE physiological effect on the limb was greatly to increase diaphoresis and the flow of blood in the skin, so that the surface had a boiled lobster colour. The most conspicuous and constant effect was anodyne. This was specially striking in painful rheumatic joints.

*   *   *   *

Well worthy of trial in painful rheumatic affections. Particulars as to the apparatus (which appears to be well worthy of trial in suitable cases, and especially for the relief of painful chronic or subacute rheumatic affections) can be obtained from the proprietors of the patent at 1 and 2, Chiswell Street, Finsbury Square, E.C.

MEETING OF BRITISH MEDICAL ASSOCIATION.

(Scotch Branches).

DEMONSTRATION.

British Medical Journal, February 8th, 1896.

Edinburgh (Fife and Lothians), Perthshire, Stirling, Kinross, and Clackmannan and Dundee and District Branches.

A COMBINED meeting of these Branches was held in the Royal Infirmary, Edinburgh, on January 31st, 1896, Dr. ARGYLL ROBERTSON, Vice-President of the Edinburgh Branch, in the chair. There was a large attendance.

*   *   *   *

Dr. JAMES showed: (1) Case of Articular Rheumatism before and after treatment by the Tallerman Localised Hot-Air Apparatus.

*   *

A demonstration of the Tallerman Localised Hot-Air Apparatus was given by Mr. TALLERMAN. Several patients were put under treatment; some of these had been similarly treated for some days in advance with lasting benefit. *In all cases shown the immediate good effected was striking.*

Demonstration by Mr. Tallerman before meeting of British Medical Association— Scotch Branches. Striking effect obtained in all cases.

10, MELVILLE CRESCENT, EDINBURGH,

July 23rd, 1896.

Dear Sir,—Since the date of my last letter I have had the opportunity of trying the hot-air treatment in some other cases, notably in peripheral neuritis and in gonorrhœal rheumatism. The results have been distinctly satisfactory.

Neuritis and gonorrhœal rheumatism: results obtained distinctly satisfactory.

In the gonorrhœal rheumatisms we were able to improve very greatly the joint mobility, and in the neuritis cases the patients' recoveries were markedly hastened.

I am glad to be able to report so favourably.

Yours faithfully,

L. A. Tallerman, Esq.   (*signed*) ALEX. JAMES.

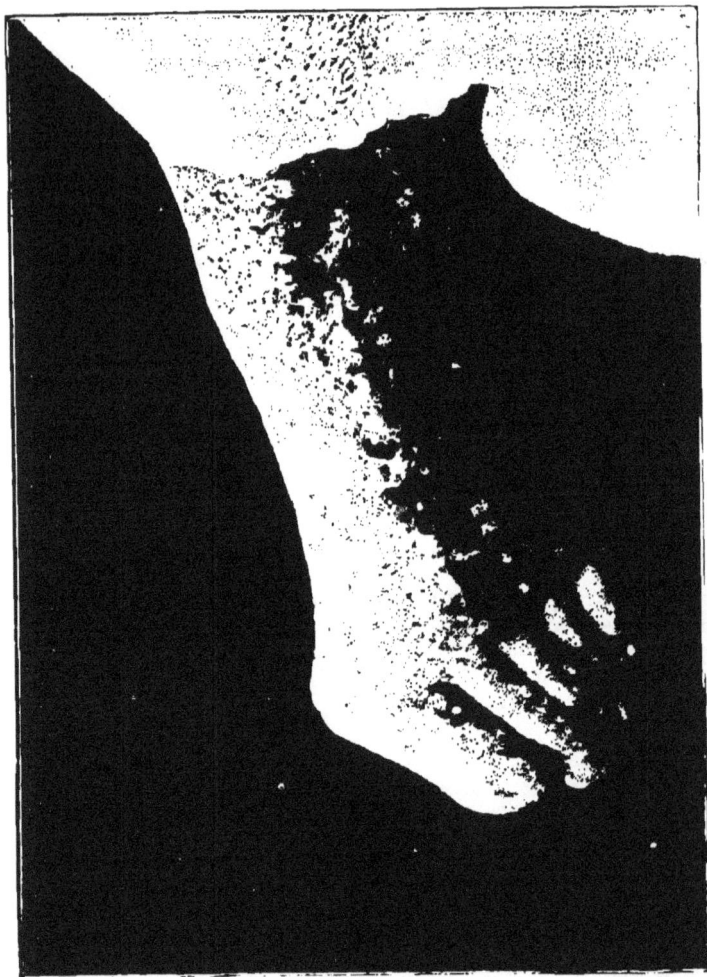

CASE 2.—Chronic Gout, 24 years, before treatment : showing enlargement and dislocation of great toe, Eczema extending up the leg, toe discharging thick pus and chalky matter, acutely painful. This left foot represents more or less patient's condition generally.

For Case notes and Report, see pages 90–93, "The Tallerman Treatment" (Baillière, Tindall, & Cox, London).

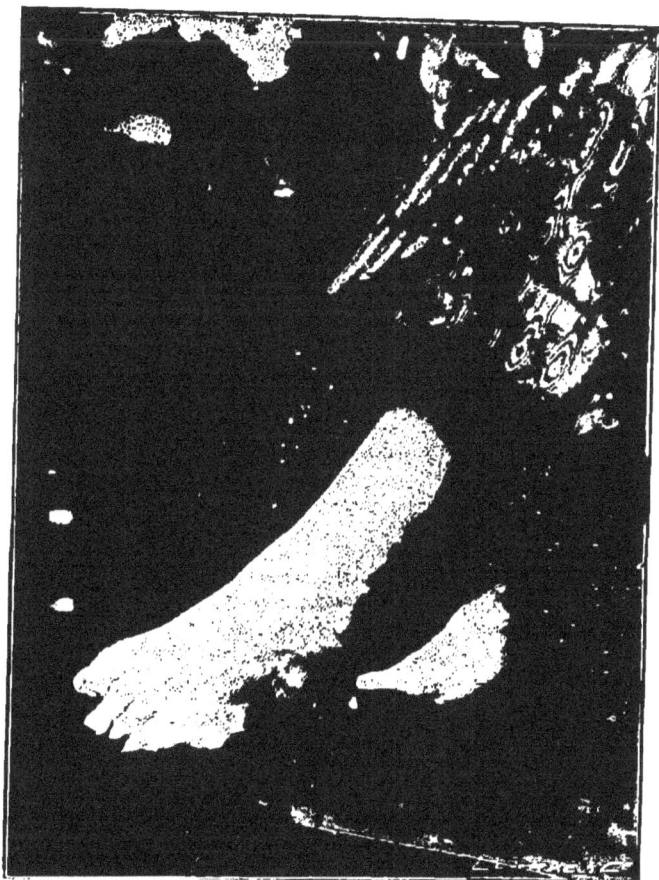

CASE 2 —After treatment: showing absence of any Eczema or Gout, and improvement in the deformity and position of the toes.

See Case Notes II., page 51.

For Case notes and Report, see page 91, "The Tallerman Treatment" (Baillière, Tindall, & Cox, London).

HARVEIAN SOCIETY OF LONDON.

Exhibition of Cases, 27th May, 1897.

Lancet, 5th June, 1897.

A Meeting of this Society was held on May 27th, the President, Dr. MILSON, being in the chair.

Treatment in phthisis, associated with rheumatism, followed by improvement.

DR. KNOWSLEY SIBLEY showed a case of phthisis associated with rheumatism and subcutaneous abscesses, probably tuberculous. The patient, a woman 64 years of age, had lost a sister from phthisis. The patient had never been very strong, but until last October, when both hands were attacked with rheumatism, she had never been laid up. Shortly after this a swelling appeared on the back of one hand and on the wrist of the other ; these had persisted ever since. The patient had an irregular temperature, usually from 100° to 101° F. at night and subnormal in the morning ; she had signs of old-standing tuberculous mischief at the right apex. Both hands were enlarged and all the fingers more or less stiff and painful ; there were considerable thickening and grating about both the wrist-joints. There was a swelling about the size of a walnut over the dorsum of the left hand and another rather larger over the ulnar surface of the right wrist. These swellings were soft, fluctuating, and not tender. The contents, drawn off with a hypodermic syringe, consisted of clear pus. A trace of albumen was present in the urine. It was unusual that a patient suffering from chronic phthisis should late in life be attacked with a sub-acute form of rheumatism, followed by what appeared to be subcutaneous abscesses probably of a tuberculous nature. Dr. Sibley did not consider the whole process to be tuberculous, as there were distinct thickening and stiffness of the individual finger-joints apart from the disease of the wrist-joints. The patient had considerably improved since she was admitted into the North-West London Hospital and had undergone treatment by the Tallerman hot-air baths.

HARVEIAN SOCIETY OF LONDON.

Two Cases Rheumatoid Arthritis treated by W. KNOWSLEY SIBLEY, M.A., M.D., B.C., Camb., M.R.C.P. Lond., by the Tallerman Method.

British Medical Journal, June 13th, 1896.

Rheumatoid arthritis with bronchitis, successfully treated.

DR. KNOWSLEY SIBLEY showed two cases of rheumatoid arthritis successfully treated by the hot-air method.

The pain in the affected or other parts was greatly relieved, and the patient experienced a considerable feeling of relief generally. Especially it was noticed that the bronchitic condition which so often accompanied this affection was also much benefited.

ANNUAL HOMŒOPATHIC CONGRESS, London,

June 28th, 1894.

Monthly Homœopathic Review, September 1st, 1894.

MR. TALLERMAN'S TREATMENT OF CHRONIC RHEUMA- *Demonstration* TISM.—The members of the Congress saw the case that was *before Annual Homœopathic* treated before them on June 28th before and after the bath, but *Congress, remarkable* those who were not present will be interested by the details of *result obtained* the case P. B., æt. 43, a painter, had acute rheumatism two *in 60 minutes in a case treated at* and a half years ago, since when he has been a patient, off and *hospital 2½ years* on, at the Royal Free Hospital without benefit to his stiff joints. *without benefit.* When he came up to the Congress meeting his fingers were swollen, painful and so stiff that he could not close either hand, though his right hand was worst. Both wrists were swollen and stiff, permitting of only very limited movements. His knee joints were also stiff and painful. His right arm was put in the cylinder, the temperature raised to 260°, and it was kept in it for an hour. At the end of this time it was released. The man was then able to close the right hand without pain, while the movements of the wrist were distinctly better. The curious thing is that the left hand and wrist, which were not treated, were also much improved. He could close his left hand without pain, and the wrist movement was also better. The knee joints were also benefited, as he said he had no pain in them on movement. This result was certainly remarkable, and the same thing—the participation of the untreated joints in the benefit has been observed in other cases. The probability is, judging by the improvement after one bath, that with a course of these local hot-air baths, this patient would completely regain the use of his stiff limbs. Another patient was brought up by Mr. Tallerman for inspection. He had been treated at St. Bartholomew's Hospital by a course of these baths.

Mr. WILLETT delivered a lecture on the subject, which was published in the *Clinical Journal.* He there gave the results of the treatment in the different cases in which it had been used, and all cases where the joints were not ankylosed were more or less benefited. This man who came up to be seen was "James L.," in Mr. Willett's lecture. His case was described as "extremely severe," one which threatened to "leave him crippled by adhesions," and "serious organic changes in the joints." The patient declared he had now no pain in any joints, the movements were quite free, and he said he was as well as ever he had been. This was certainly a remarkable and noteworthy result.

This mode of treatment for chronic rheumatism and rheumatoid arthritis promises to be a very valuable addition to our means of cure.

F

BRISTOL MEDICO-CHIRURGICAL SOCIETY.

British Medical Journal, December 4th, 1897.

. . . . Dr. E. C. WILLIAMS showed the following cases. treated by the localised application of superheated dry air: (*a*). Monarticular Rheumatoid Arthritis, (*b*) Polyarticular Rheumatoid Arthritis. Dr. JAMES SWAIN made some remarks.

THE LANCET, August 29th, 1896.

The Tallerman Treatment in Rheumatism and Allied Affections,. by W. KNOWSLEY SIBLEY, M.A., M.D., B.C. Camb., M.R.C.P. Lond., Senior Physician to Out-Patients at the North-West London Hospital.

THE dry hot-air method, was first brought under my notice in August, 1894, when an apparatus was supplied to the North-West London Hospital and left for the use of the staff. I have tried the treatment in a large variety of cases ever since, and now wish to publish some of the results. This period of two years has enabled me to form a definite opinion that the value of the treatment is not merely a temporary one, but is of a more or less permanent nature.

*   *   *   *

Pysiological effect produced bitherto considered impossible. When the treatment is required to act quickly as an anodyne the temperature is rapidly raised to 260° or 280° F. But under ordinary circumstances such as those described below, it is gradually raised and a general free perspiration breaks out over the whole body; at the same time the body temperature is temporarily raised from a half to three degrees, a physiological effect hitherto regarded as impossible to be obtained. Also the pulse increases in frequency and to a less marked extent the respiration. A few minutes after the operation is completed the pulse, respiration and temperature, return to the normal or previous condition. Usually about an hour after the pulse is found to be slower and stronger than it was before treatment; this was especially noticed in some cases of weak and enfeebled hearts. In cases accompanied with much pain this is almost at once relieved, and under the influence of the heat the parts soon become more lax and supple. When the limb is first removed there is often a transient erythematous blush.

*   *   *   *

Portable character of apparatus enables the treatment of patients in their own houses. The portable character of the apparatus enables it to be taken to the sick room and used by the bedside in cases where it would be impossible to move the patient. The local bath gives far more successful results as a method of treatment.

The following cases are briefly described in the order in which they came first under my treatment at the hospital, not necessarily in the order in which they were subjected to the hot-air method. Many were for some time previously treated by the ordinary drug routine, and only after failure by this were they submitted to the new one. If there has been any selection of cases it will be at once apparent that it has been on account of their chronicity or severity. Most of the notes are taken from my hospital notebook made on the dates as reproduced and with regard to symptoms more or less in the patient's own words.*

* * *

CASE I.—*Arthritis deformans; duration four-and-a-half years.*—A single woman aged sixty-four years. Has been treated at King's College Hosital and North-West London Hospital, getting gradually worse.

> August 10th, 1894. Could not feed or dress herself—first treated.
>
> August 24th. Returned to work—had been treated nine times.
>
> September 9th. Considered herself practically cured—treated twenty times.

* * * *

CASE II.—*Sub-acute articular rheumatism; aortic regurgitation; psoriasis.*—Married woman aged nineteen years.

> October 4th, 1896. Treatment commenced.
>
> October 25th. Pain nearly all gone, able to run to catch a train, which she had not done for months.
>
> November 11th. There had been no more rheumatism. Treated six times.

* * * *

CASE III.—*Chronic rheumatism; duration eight years.*—An unmarried woman aged sixty-one years.

> October 30th. Unable to raise her arms, lost her work.
>
> November 11th. Shoulder freer. Had been treated twice.

* *

---

* For full Notes and Report upon these cases, see *Lancet* 29th August, 1896; also "The Tallerman Treatment," edited by Dr. S. Shadwell, M.A., M.B. Oxon., M.R.C.P. London, with 63 photographs. Baillière, Tindall, & Cox.

CASE IV.—*Arthritis deformans ; duration eight years.*—
Married woman aged sixty-nine years.

> April 22nd. Knees and ankles stiff and swollen, got
> about with difficulty, hands greatly deformed, fingers
> enlarged and very painful, had not been able to use
> them for anything for eight months though under
> medical treatment all the time.

See Photos—before, during, and after treatment, pages 34 and 35

CASE V.—*Arthritis deformans ; duration thirty years.*—
A woman aged fifty-one years.

> June 19th. Hands, shoulders and knees swollen and
> painful, and getting worse, unable to raise arms or
> close hands, dragged herself about with great
> difficulty.

> July 10th. Treated ten times, quite comfortable, except-
> ing knee, owing to patient living at top of house and
> having to mount several flights of steps.

CASE VI —*Chronic rheumatism ; duration ten years.*—An
unmarried woman aged fifty-nine years.

> July 20th, 1896. Suffered constantly from rheumatism,
> severely in legs and feet ; after first treatment pain
> gone.

> July 27th. Third treatment, walked better than for years
> contraction of toes disappeared

> July 30th. Fourth treatment, reported herself well.

CASE VII.—*Chronic rheumatism.*—Married woman aged
sixty-one years.

> July 21st, 1896. Left shoulder stiff, can raise arm very
> little, hands and fingers swollen and painful, marked
> wasting of muscle, becoming worse, pulse feeble and
> small. After first treatment could extend arm, fully
> close hand.

> July 27th. Treated four times, much better, could do a
> little work.

> August 6th. Treated six times. All movements of
> fingers, hands and arms quite free. Still a little
> numbness in left arm, sleeps well.

CASE VIII.—*Sub-acute rheumatism; mitral regurgitation.*—
A married woman aged thirty-two years.

> July 21st. Could not raise arm or close hand, little sleep
> at night for more than a week, depressed. After first
> treatment extended arm fully over head, hands
> clenched, pain and depression gone, walked home,
> unable to do this for six months.

> July 30th. After third treatment pain all gone, except a
> little in left shoulder.

* *

CASE IX.—*Lumbago and sciatica; duration six weeks.*—
A man aged twenty-six years.

> July 20th. Had sciatica and lumbago for six weeks, after
> treatment pain much better.

> July 21st. Second treatment. Complete and satisfactory
> result.

> July 23rd and August 6th. No pain since last treatment,
> patient had returned to work.

* * * *

CASE X.—*Chronic rheumatism; duration six months.*—
Man fifty-nine years of age.

> July 16th, 1896. Persistent pain right shoulder, worst at
> night, and unable to work, could not raise arm above
> horizontal.

> July 22nd. Treated three times, since less pain, able to
> do some work, slept much better.

* * * *

CASE XI.—*Sciatica; duration seven months.*—A man aged
fifty-six years.

> July 16th. Becoming worse and prevented him working,
> walked a little with pain and difficulty, could only
> raise leg a few inches from ground.

> July 17th. Patient had walked nearly five miles for
> second treatment without much difficulty.

> July 20th. Third treatment and last. On 28th July
> reported he had no pain and walked without a stick,
> which he had not done for six months.

See Photos—pages 74 and 75.

* * * *

CASE XII.—*Neuralgia following herpes.*—A woman aged sixty-five years.

> July 20th. Severe neuralgic pain head and neck, could not sleep, depressed and crying, after forty minutes all pain gone, some return at evening.
>
> July 29th. Treated second time, pain again disappeared
>
> July 30th. Little or no pain hip or thigh, but little in lower part of leg. July 31st, reported all pain had gone.

*   *   *

CASE XIII.—*Sciatica.*—A man aged fifty-one years.

> July 28th, 1896. Had had sharp attacks of sciatica, after a week had to give up work, rubbed the part with embrocation, took the skin off, no relief.
>
> July 30th. First treatment, all pain had gone from hip and thigh, but some in lower part of leg.
>
> July 31st. Third treatment, all pain quite disappeared.

*   *   *

It must be admitted by those who have had much practical experience of severe cases of arthritis deformans such as Cases Nos. I., IV. and V., how very hopeless these have generally been considered and what little chance of permanent good medical treatment offers, that this treatment by hot air at a very high temperature meets a general want. I must add I have never seen results so immediate and satisfactory produced by any other treatment. It is now two years since treatment by the local hot-air bath was commenced in the first case, and yet the patient continues comparatively free from the complaint, and even the deformity of the fingers has greatly disappeared. Very intractable cases of sciatica, such as Case VII., usually pass into the hands of the surgeon, who performs nerve stretching, often with very little good result. It seems likely that these cases will likewise in future be cured by this less drastic means. Another possibility of much importance may be the prevention of many cases of the morphia habit. Judging from the general physical improvement that all of the patients showed after undergoing treatment, it would appear that this method will be found beneficial for cases of chronic bronchitis and for some cases of chronic heart mischief. Many of the patients have stated that their bronchitis was better than it had been for years. I possess also evidence of the undoubted value of this hot-air treatment in acute and chronic gout and hope shortly to publish a series of cases.

Marginal notes: Never saw results so immediate and satisfactory produced by any other treatment. Two years since treatment in first case was commenced and patient continues free from the complaint.

Marginal note: Bronchitis improved.

*   *   *

The treatment appears to lower the blood pressure of the body, and in some way to increase the alkalinity of the blood which enables it to dissolve the uric acid from the tissues and joints and get rid of this substance through the various excretory organs. This is evidenced by the relief from local pain and the removal of the frequent uric acid nerve depression. Hence the treatment is of a tonic nature and bestows an increased general vitality upon the patient.

*   *   *   *

Undoubtedly the writer's experience points to the conclusion that the chronicity of a case is by no means a bar to a successful issue. With regard to rheumatoid arthritis many forms of rheumatism with chronic joint mischief, sciatica, lumbago, and one might add some cases of neuralgia, neuritis, and chronic bronchitis, there can be no question as to the great benefit of the treatment. The record of the remarkable results previously published obtained by this apparatus at the North-West London Hospital alone embraces such a wide field that this localised hot-air bath must be reckoned upon in the future to play an important part in the relief of pain and the cure of disease. Especially is it likely to prove of great use in those forms of very chronic disease which have hitherto yielded but little to any known medicine, the sufferers from which are commonly sent in search of continental hydrothermic establishments, and usually, sooner or later, fall into the hands of the quack.

Chronicity of a case no bar to successful issue. No question as to the great benefit in—Rheumatoid arthritis, Rheumatism, Chronic gout, Sciatica, Lumbago, Neuralgia, Neuritis and Chronic bronchitis.

MEETING NORTH-WEST LONDON CLINICAL SOCIETY.

Demonstration of Cases.

Dr. CUBITT LUCY in the Chair, March 17th, 1897.

Clinical Journal, April 28th, 1897.

Dr. KNOWSLEY SIBLEY showed three cases treated by the Tallerman dry hot-air method.

*   *   *

THE cases he showed illustrated the improved condition in the nutrition of the parts so treated. The first case was one of chronic ulcer on the leg of a man æt. 60, who came under his care suffering from general bronchitis and dyspepsia. The ulcers—two in number—had existed more than a year, and were situated just below the left internal malleolus, being about one and a half inches in diameter. There was a good deal of general œdema of the part, and the leg was very painful. The patient was ordered some boracic ointment; and as the ulcers were getting larger instead of smaller he was ordered the hot-air baths, the first of which he had on February 2nd. After two

Chronic ulcer, improved nutrition under this treatment.

baths it was obvious that the ulcers were healing, and the healing was complete on March 1st, by which time the patient had had ten baths.   The general nutrition of the whole limb had also considerably improved, there was less œdema, and the inflammatory area around the ulcers had disappeared.   He considered that this hot-air treatment compared very favourably with the oxygen treatment for ulcers and other diseases.   The second patient had suffered from malnutrition of the leg after ligature of the femoral artery and injury to the nerve.   He was in the hospital under Mr. Durham from September, 1895, to July, 1896, with an abscess in the popliteal space which eroded into the popliteal artery, the leg was cold, very blue, and much withered.   There was considerable impairment of sensation in places, especially about the foot, complete anæsthesia of the great toe and inner side of the foot; there was also perversion of sensation, the ankle was very stiff, and the patient could hardly walk.   By March 15th the patient had had eighteen baths.   The limb was much stronger, and he was able to get about comfortably; the feeling in the foot was much improved, and the limb remained constantly warm.   The general nutrition of the part had also improved.   The third case was one of malnutrition of the hand due to severance of the nerves of the wrist.   The patient received a severe cut across the wrist-joint in October last year, which was followed by loss of sensation in the last three fingers, and the hand could not be used for anything.   On February 16th he had had six baths, and was able to use his hand a little.   Sensation had also returned to a considerable degree in the fingers.

Compares very favourably with the oxygen treatment.

Mr. JACKSON CLARKE agreed that, like massage, hot-air baths had the property of improving the nutrition of particular parts.   One of the patients exhibited he noticed had some degree of equinus, and he recommended a small operation being done when the limb was sufficiently well nourished for the patient's future comfort.   The hot-air treatment was a valuable adjuvant to surgical treatment.

A valuable adjuvant to surgical treatment.

MEETING OF THE NORTH-WEST LONDON CLINICAL SOCIETY.

Demonstration of Cases.   Dr. McEVOY in the chair.

Clinical Journal, April 6th, 1898.

THE TALLERMAN TREATMENT.

DR. KNOWSLEY SIBLEY showed a case of Bright's disease associated with asthma and bronchitis.   A man fifty-eight years of age, employed in a horseyard, had suffered from asthma on and off for twenty years, especially in the winter ; for the last three years he had also had a good deal of bronchitis.   Two

Bright's disease with asthma and bronchitis. Laboured breathing improved. Ability to sleep. Only occasional trace of albumen.

years ago he had dropsy of the legs, which lasted for five months; this was probably associated with albuminuria. He was in University College Hospital in September, 1895, for eight weeks, and again in January, 1896, for three weeks with the present complaint. Patient first attended as an out-patient at the North-West London Hospital at the beginning of April, 1896. He then had frequent attacks of asthma, the pulse was very firm, and there was a good deal of albumen in the urine. He attended regularly from this time until November, and the following drugs were in turn prescribed, but without giving him any real relief: turpentine, iodide of potassium, ipecacuanha, senega acids and stramonium. On November 26th the following note of his condition was made:—"Patient had a very bad attack again last night, with great choking and dyspnœa. For some time it has been impossible for him to lie back in bed. There are a few rhonchi, especially on the right side; the second cardiac sound is much accentuated, the first sound not clear." On November 30th he was first treated by the Tallerman dry hot-air method, the right arm being placed in the apparatus. The breathing, which was much laboured before, was greatly relieved by the bath, and although no change was made in the medicine he had a very much better night. On December 7th he had his fourth bath, and the nurse made the following note:—"Slept very well after last bath, seems much better; says he has not felt so well for months." On December 17th Dr. Sibley made the following note:—"Patient has had seven baths, is better than he has been at this season of the year for many years, very little rhonchi left. Baths were discontinued at the end of January, 1897, he having then had fifteen. By the middle of February, 1897, he was able to sleep for seven or eight hours together; at the beginning of November he went to a convalescent home for three weeks, which did him good. On December 9th he came to the hospital again, the cough having become much worse, and he was nearly choked with the phlegm; after one or two hours' sleep he would, every night, wake up in great distress, and be unable to get to sleep again for some hours. The pulse was very firm, and rate 80; there were scattered rhonchi over the lungs, with very prolonged expiration. There was a small quantity of albumen in the urine. On December 13th he recommenced the baths, with the result that he slept the whole night afterwards, and felt much better the following day. The baths have been continued since at intervals, with the result of great improvement in every way. For some days the albumen quite disappeared from the urine, but it now reveals a slight trace again occasionally.

Dr. Harry Campbell congratulated Dr. Sibley on the good result he had obtained. The question arose whether the patient was better than he would have been by the ordinary hot-air baths. ✻ ✻ ✻ ✻

Dr. Guthrie asked what effect the hot baths had upon the secretion of urine. Personally he was not fond of hot baths in

CASE 5.—Sciatica (taken July 16th, 1896), seven months and becoming worse, before treatment : showing limit of patient's ability to move. This is the highest point to which the limb could be raised, and the movement was accompanied by pain.

For Case notes and Report, see page 20, " The Tallerman Treatment (Baillière, Tindall, & Cox, London).

CASE 5 —Sciatica (taken July 20th, 1896) showing patient's ability to move the leg after the third treatment. Had walked five miles and without stick, which he had not done for six months.

For Case notes and Report, see page 20, "The Tallerman Treatment" (Baillière, Tindall, & Cox, London)

nephritis, because there was always a certain risk ; the perspiration reduced the amount of water available for excretion by the kidneys, and also the amount of poisonous substances excreted by the skin was very much less than that given off by the kidneys, so that there was danger of the accumulation of these substances on account of the smaller quantity of water passed by the kidneys.

Dr. Sibley, in reply, said . . . . he had a number of patients treated by the Tallerman baths who had previously had the ordinary hot baths without benefit, but they improved distinctly under the Tallerman system.

❖

Increased elimination of uric acid.

He was coming to the conclusion that when the patient had had a few baths the quantity of urine secreted was increased, notwithstanding the profuse perspiration, and the uric acid and urea excreted increased as time went on. At the time of starting the baths the relation of uric acid to urea was often 1 to 33 and 1 to 40, and after a few baths it rose to 1 to 17 and 1 to 20. Some

No drugs given to patients under treatment at the Laennec Hospital.

interesting observations with the Tallerman method were made a short time ago in one of the French hospitals, and the results were particulary valuable because the patients were not having any drugs internally. "La Presse Médicale" published these cases, which were treated at the Laennec Hospital, Paris, under Professor Landouzy and Dr. Oulmont.

MEETING PHILADELPHIA COUNTY MEDICAL SOCIETY.

Philadelphia Policlinic, November 14th, 1896.

Wonderful improvement in lumbago and lead poisoning.

THERE was also exhibited at the same meeting a Tallerman apparatus for local hot-air baths, and two patients, one with acute lumbago, the other with chronic saturnine gout, were treated by one-half hour's application of dry heated air at a temperature averaging 248° F., 260° being the maximum. Both patients were much improved, indeed, wonderfully so.

AMERICAN ORTHOPEDIC ASSOCIATION.

Treatment of Rigid Degrees of Flat-foot by the Tallerman Localized Hot-Air Bath. Paper by W. J. WALSHAM, F.R.C.S. London, England (*Transactions viii.*, 1895).

"FOR the last fourteen years I have been accustomed to treat "the numerous cases of rigid Flat-foot that came under my care

"in the Orthopedic Department of St. Bartholomew's Hospital, Adoption of the treatment for rigid flat-foot.
" either by manipulation and massage, or by wrenching under
" an anæsthetic. For the last six months or so I have substituted
" for this treatment the use of the Tallerman hot-air bath. The
" class of cases most suitable for manipulation and massage are
" those in which on taking the foot in two hands and making
" pressure on the head of the astragalus with the ball of the
" thumb, whilst at the same time adducting and inverting the
" front of the foot, the muscles gradually yield, and the foot can
" be made to assume its normal shape, the arch being completely
" restored for the time being. This manipulation, however, is
" attended with acute pain, and on relaxing the pressure the
" foot at once resumes the deformed position. By repeating the
" manipulation, which on each application becomes less painful,
" daily, or better, several times a day, for a week or two, the
" rigidity can generally be overcome, especially if the patient
" can take rest during the treatment. The foot is then in a
" condition for the flattening to be permanently cured or relieved
" by exercise, combined with some form of valgus boot. For
" this variety of rigid flat-foot the hot-air bath is most useful.
" After the rigid foot has been in the bath at a temperature
" varying from 270° to 280° F., or higher, for half-an-hour to
" three-quarters-of-an-hour, it comes out quite supple, and on
" taking it in the hand the arch can be restored without any
" pain or the application of any force. After a few hours the
" rigidity gradually returns, and the process, like the manipulation,
" has to be repeated. The bath does not, of course, cure the
" flat-foot, but merely by getting rid of the rigidity places it in a
" condition to be acted upon by the exercises or other means
" employed for its permanent relief. The advantage of the bath
" treatment over the manipulative consists in the entire absence
" of pain attending it."—(*For cases and further observations see
Paper, vol. viii., Association's Transactions.*)

WALSHAM'S "THEORY & PRACTICE OF SURGERY."

London, 1897.

Chronic Rheumatic Arthritis (Treatment of) —— . . . Of service for softening adhesions and restoring movement.
" I have found the Tallerman localized hot-air bath of service in
" these cases for softening adhesions and restoring movement."

LIVERPOOL WORKHOUSE COMMITTEE MEETING.

Liverpool Daily Post, Sept. 18th, 1896.

The fortnightly meeting of the Workhouse Committee was held
yesterday, Mr. J. Hodgkinson in the chair.

A QUESTION arose as to the hiring of a (Tallerman) patent localized
hot-air apparatus for the cure of rheumatism. Dr. Alexander,

Thirty cases of chronic rheumatism relieved. medical officer of the workhouse, reported that the apparatus had been on trial for two months, and had relieved thirty cases he previously considered as chronic.

It was decided to accept the offer of the patentees to let the machine on hire.

CHELTENHAM BOARD OF GUARDIANS' MEETING.

Cheltenham Examiner, 9th June, 1898.

The weekly meeting of the Board of Guardians was held in the workhouse board-room on Tuesday, Mr. W. F. Hicks-Beach in the chair.

A GOUT AND RHEUMATISM CURE.

THE medical committee recommended that a Tallerman apparatus for the treatment of gout and rheumatism should be hired for one year, and that the workhouse medical officer (Dr. Pruen) should be authorized to spend a sum on the fittings.

In moving the adoption of the report, Mr. Carrington explained that the question of the provision of the apparatus referred to had been adjourned from a previous meeting in order that the committee might inspect it. This they had now done, and they considered it was very desirable to have the apparatus, which, however, was not at present in the market for sale, but could be hired. It had been employed very effectually in curing gouty and rheumatic cases, and as in one instance already the result of its application had been to enable a recipient of out-door relief to return to work, there was reason to believe that in this way the cost of having it would be balanced by saving relief.

Out-door relief saved by the use of the Tallerman treatment. Mr. Bence pointed out that they were not incurring the expense without warrant, because they had the fact that through the use of the apparatus they had got rid of an out-door relief case to which they had been paying 5s. per week.

The report was adopted.

CHELTENHAM BOARD OF GUARDIANS.

Cheltenham Examiner, August 24th, 1898.

THE TALLERMAN PROCESS.

A REPORT from the Medical Committee, based on reports from Mr. Braine-Hartnell (Dr. Pruen's deputy), stated that ten cases of rheumatism and allied affections had been under treatment, and all, except one case, had benefited, most being " distinctly better."

Cheltenham Chronicle, August 20th, 1898.

THE Tallerman apparatus for the cure of rheumatism and kindred disorders is, however, getting bold advertisement in Cheltenham (for which I hope Mr. Tallerman is duly grateful) by the fact that the Guardians are giving it a public trial. They have, up to the present, tried it on ten paupers, with the result *Ten paupers treated, nine* that nine of them are distinctly better. Nevertheless it must be *distinctly better.* an extremely trying ordeal for our rheumatic friends, as some have had to take twenty, some fifteen, and some ten baths in the course of one short month. If cleanliness be next to godliness, Tallerman's system has certainly much to recommend it on that ground alone ; but I am sure that at least one or two of the old ladies and gentlemen, judging from the antipathy to water frequently observed amongst pensioners on the poor rate, must feel uncomfortably clean.

Lancet, November 13th, 1896.

TALLERMAN TREATMENT.

To the Editors of *The Lancet.*

* * * *

THERE is nothing new then in the local application of hot air. The peculiarity of the Tallerman treatment consists in the use of such high temperatures as 300° F., which is only possible by keeping the air dry. In an ordinary box the perspiration, which is soon excited, shortly renders the atmosphere so moist that temperatures of 110° and 120° F. are as much as can be borne with comfort. In short, with ordinary apparatus a local hot-air bath is nothing more than a vapour bath, and all those who are practically acquainted with the question will recognise the value of the Tallerman invention if it really provides us with a means of keeping the hot air dry and reaching a temperature of 300°.

I am, Sirs, yours truly,

Paris, November 3rd, 1896.   OSCAR JENNINGS.

LANCET, October 17th, 1896.
LOCAL HOT-AIR TREATMENT.

To the Editors of *The Lancet.*

SIRS,—Referring to the report of cases by Dr. W. K. Sibley in *Chronic sciatica and lumbago.* *The Lancet* of August 29th last, which were treated by the above *Relief sought by* method, I am glad to be able to confirm from personal *physician to three hydros.* experience the curative effect of a course of only four baths for *Treated four times, is free from* chronic sciatica and lumbago, which had begun to extend to the *pain, and walked three miles.*

sound leg. In the severe winter of 1894 I had an attack of acute sciatica which has continued in a chronic form. Having acted as resident physician to three hydropathic establishments I had enjoyed considerable experience in the treatment of such cases, and was enabled to apply the knowledge so acquired to my own case. The relief obtained was but slight and I had difficulty in getting about, the pain at times being very severe. I am glad to be able to record that I derived benefit from the first hot-air bath of the Tallerman process. Having now taken four I am free from pain and able to walk three miles at a stretch. No ill effects or risk of any sort are likely to attend a course of these baths, a slight rise of temperature and increase of pulse being alone indicated. The baths were taken at the institution, 50, Welbeck Street, W.

I am, Sirs, yours faithfully,

F. FITZHERBERT JAY, M.D. St. And.

Constitutional Club, October 7th, 1896.

THE LANCET, May 7th, 1898.

The Tallerman Treatment by Superheated Dry Air: Case Notes and Medical Reports, with numerous Illustrations. Edited by ARTHUR SHADWELL, M.A., M.B. Oxon., M.R.C.P. Lond.   London: Baillière, Tindall, and Cox. 8vo, pp. xii. and 173.   63 Plates.   1898.

Failure of previous attempts. MANY endeavours have from time to time been made to utilise high temperatures in the treatment of disease, and in several instances a certain measure of success has followed these attempts ; but the temperature attained by these methods has never reached any very great height and, in fact, in no case has it approached the boiling point of water.   The failure we now know was due to the fact that the heat was applied either by means of steam or, if dry air were at first used, it became very quickly saturated with moisture by the increased perspiration of the part treated, so that practically steam was applied in this way also.

Value of treatment in rheumatism and rheumatoid arthritis. The method is undoubtedly of value in several forms of joint disease, especially in chronic rheumatism and rheumatoid arthritis which are so little amenable to ordinary treatment, and in this book many cases are recorded in which other diseases, such as peripheral neuritis, chronic ulcers, and flat-foot have been benefited. The treatment has been employed fairly extensively, but is certainly deserving of being more widely known. Of course, it is not a panacea and time will enable us to distinguish what cases are likely to derive benefit from it. The power of relieving pain is sometimes very great, and the increased freedom of movement in chronic joint disease is equally remarkable. The book contains notes of a large number of cases in which the apparatus has been employed, and in many instances photographs are given of the patients before and after the treatment. Mr. Tallerman is to be greatly commended in that he has not endeavoured to bring his in- Recognition and vention before the public but has confined its employment to approval of the inventor's the members of the medical profession. We feel sure that attitude towards the profession before long the Tallerman method of applying air at high recognized and temperatures to diseased portions of the body will take the place approved. High commendation of it deserves in the estimation of medical men as the most satis- the Tallerman treatment. factory method at their disposal of treating many hitherto very intractable morbid conditions.

MEDICAL PRESS, June 8th, 1898.

SHADWELL ON TALLERMAN TREATMENT.

THIS work forms a monograph upon the Tallerman treatment of various conditions, chiefly of a rheumatic, rheumatoid, or gouty nature, by superheated dry air. It is the work of a writer who knows how to present his facts to the reader in a clear light, and will no doubt form a book of reference for the particular branch of practical therapeutics with which it deals.

Mr. Shadwell has wisely devoted the greater portion of his book to the careful record of clinical cases, which are abundantly illustrated with full-page blocks in evidence of such success as the unlocking of joints, and the subsidence of swellings following the treatment. Some of his most striking records are those dealing with the treatment of rheumatoid arthritis. Every practical physician must have experienced at one time or another a feeling of hopeless despair when called Despair of physicians when upon to treat that terrible and progressive disease. But if we called upon to treat rheumatoid are to believe the testimony of this book, cases of four and five arthritis. years' standing, with implication of numerous joints, have been practically cured. Several observations are of great interest— for instance, although only one part of the body be under treat- ment, yet improvement may follow in other and remote joints. The local phenomena, *e.g.*, reddening and sweating, are attended by general disturbances, such as raised body temperature and quickened pulse. One noteworthy feature is the instant relief

G

No medical man *of* rheumatic and rheumatoid pain, and were there no other
can do justice to good result to chronicle, that alone would more than justify the
his patients in
gout, rheumatic, appearance of the present work. It may safely be said that *no*
or rheumatoid
types of disease *one who undertakes the treatment of chronic diseases of the gouty*
unless he has *and rheumatic or rheumatoid types can expect to do full justice*
studied the
claims of the *either to his patients or to himself unless he has studied the claims*
Tallerman's *of the Tallerman treatment.*
treatment.

DUBLIN JOURNAL OF MEDICAL SCIENCE.

March, 1898.

The Tallerman Treatment by Superheated Dry Air in Rheu-
matism, Gout, Rheumatic Arthritis, Stiff and Painful Joints,
Sprains, Sciatica, and other affections. Case Notes and
Medical Reports, with numerous Illustrations. Edited by
Arthur Shadwell, M.A., M.B. Oxon., M.R.C.P London :
Baillière, Tindall, & Cox. 1898.

" THIS book presents in a convenient form full and authoritative
" information respecting the Tallerman treatment, which, the
" preface tells us, is still very imperfectly known.

" That hot, dry air was a powerful therapeutic agent has
" long been known, and for ages the Turkish bath was the
" nearest approach made towards the desired end of using hot,
" dry air.

" Mr. Tallerman some five or six years ago succeeded in
" producing an apparatus by means of which hot, dry air could
" be applied to any part of the exterior of the body. A descrip-
" tion of the apparatus is given, and its method of application
" is told in the book.

" Good grounds exist for the belief that, properly applied,
" hot air can affect diseased tissues beneficially, and the related
" experiences of competent and trustworthy members of the
" medical profession go to show that as a therapeutic agent the
" value of hot, dry air has passed from the theoretical to the
" practical stage."

MONTHLY HOMŒOPATHIC REVIEW. April 1st, 1898.

. . . . The testimony of this volume* is certainly considerable
and weighty, not the least part of which is found in the following
lengthy quotation from the preface by Dr. Shadwell. He

---

* Dr. Shadwell on the Tallerman Treatment. Baillière, Tindall, & Cox, London.

writes : " When requested to supervise the preparation of the volume I readily consented, for three reasons. In the first place, experience has convinced me of the value of the treatment ; in the second, I think it ought to be very much better known than it is ; and, in the third, I have no personal interest whatever. I originally approached Mr. Tallerman's invention with the scepticism which becomes second nature to a medical man, but having tested it on my own corpus vile, I found that it did what it pretended to do ; and then I saw a boy with a knee-joint full of fluid and wincing at every movement, gradually charmed off, within half-an-hour, into a smiling and painless indifference, which permitted the free handling and flexion of the knee without a murmur. Since then I have repeatedly seen results produced in old and hopeless cases of rheumatic arthritis which I could not have believed on any lesser evidence than my own eyesight. The facts related in this volume amply corroborate The value of the treatment my experience, and make it unnecessary for me to say anything testified to by more on that head. Attested as they are by many independent evidence no one can affect to observers of high standing in the profession, they form a body ignore. of evidence which no one can affect to ignore or despise. They do not come from one or from a few cliniques, but from a large number of first-rate hospitals, not only in this country, but in Paris, the United States, and Canada. It is impossible to deny the weight of so large a mass of concurrent testimony."

WEST LONDON MEDICO-CHIRURGICAL JOURNAL,

April, 1898.

THE TALLERMAN TREATMENT.

IN this book are grouped the various diseases which have been submitted to the above treatment; clinical notes by qualified practitioners, which have appeared in the journals, have been brought together, thus enabling the reader to form an opinion as to the merits of the invention.

The success that has followed the use of superheated dry-air baths is specially noteworthy in cases of rheumatoid arthritis and gonorrhœal rheumatism, which are proverbially rebellious to ordinary therapeutic measures.

Stiffness of joints after tuberculous and other inflammatory conditions are among the most troublesome cases we have to deal with, and many such are reported as having been cured or remarkably relieved by the Tallerman treatment.

Case 1., p. 24, is certainly one to be proud of, and it deserves A case to be proud of. the number of photographs with which it is illustrated. Case 19, p. 35, is that of a pianist who had rheumatic arthritis to such an extent that he was unable to follow his profession after three years' ineffectual treatment by drugs and visits to Bath and Aix. He had the superheated dry-air baths, and in three months he was able to accept professional engagements.

It is not claimed that the treatment has any marked·
beneficial effect upon cases of organic heart disease, though some of these have been temporarily relieved ; but it is important to note that the existence of cardiac complications is no bar to the use of the baths.

*   *   *

The book is well got up, is well supplied with photographic illustrations, and undoubtedly fulfils its purpose of describing with all the necessary detail the method of the Tallerman treatment, about which information of late has been frequently asked for.

TREATMENT, March 24th, 1898.

DR. SHADWELL has arranged his material under the headings ot rheumatic arthritis, rheumatism (acute and chronic), gonorrhœal rheumatism, inflamed joints, gout, sciatica and allied affections, sprains, chronic ulcers, and flat-foot. Under each head well-authenticated cases are described, and where possible photographs and skiagraphs are given of the condition of the patients and joints before and after treatment. A body of evidence is thus brought together which tends to show that the Tallerman treatment by perfectly dry air at a high temperature is serviceable in many cases for the cure of disease and the relief of pain. Messrs. Baillière, Tindall, & Cox have evidently spared no expense to render the book attractive. We welcome it as a clear explanation of what promises to be a very useful addition to the treatment of a troublesome class of chronic cases.

D'A. P.

MEDICAL MAGAZINE, March, 1898.

THE treatment of joint affections in the "superheated dry-air cylinder," invented by Mr. Lewis A. Tallerman, has been practised in this country and elsewhere during the last three or four years.

*   *   *   *

We are able from our own observation to confirm the statement that a profound thermal effect is produced by this means. The activity of the circulation is greatly heightened, the calibre of the arteries increased, free diaphoresis sets in, the pulse rate is accelerated and the body temperature raised. The immediate effects on the limb are usually seen in relaxation of the tissues and increased mobility of the joints and muscles. This is particularly obvious in cases of stiffness from muscular spasm. There is also in favourable cases relief of pain and swelling if these are present.

The trials made by Mr. Alfred Willett and others with Mr. Tallerman's special form of hot-air apparatus, encourage the belief that superheated air may prove a valuable stimulant to nutrition in damaged tissues. *Valuable stimulant to nutrition in damaged tissues.*

HOSPITAL, March 19th, 1898.

The Tallerman Treatment by Superheated Dry Air. Case Notes and Medical Reports, with Numerous Illustrations. Edited by ARTHUR SHADWELL, M.A., M.B. Oxon., M.R.C.P. (London : Baillière, Tindall, and Cox, 1898).

THIS is a book which should be read by all medical men who wish to inform themselves as to what can be done by a method which is somewhat outside the ordinary medical routine, for the relief of rheumatism, gout, rheumatoid arthritis, and other affections of joints and fascia attended with pain and stiffness. *A powerful means of treating a most obstinate class of disease.* The Tallerman treatment is described as a new method of applying heat locally to the cure of disease and the relief of pain, its most important feature being the use of dry air, which allows a much higher temperature to be applied to the part affected than would otherwise be possible. Hot water, as is well known, becomes painful at about 115° Fahr., and vapour or steam cannot be borne above 120° Fahr., whereas it has been ascertained that hot air, when dry, can be tolerated up to 300° Fahr., and even higher. The book contains many careful observations made by Dr. Sibley in regard to the effect of the local application of hot air according to this method on the general temperature, the pulse, and the respiration. The main part of the volume consists of records, in many instances admirably illustrated, of cases which have derived benefit from this treatment. The cases dealt with are principally rheumatic arthritis, rheumatism, stiff joints, sciatica, and sprains ; but an interesting paper by Mr. Walsham is included, giving his experience of the method in the treatment of flat-foot. It is impossible to peruse this book without seeing how powerful a means of treating a most obstinate class of diseases is placed at our disposal. Altogether, it is an interesting clinical record.

NORTH-WEST LONDON HOSPITAL.

Medical Press, August 18th, 1897.

THE TALLERMAN TREATMENT.

The Tallerman treatment has lent valuable assistance in relieving the pain, suffering and distress amongst a large class of patients visiting the N.-W. L. Hospital.

SIR, — Will you allow me to acknowledge through the medium of your columns a gift by Mr. Lewis A. Tallerman to this hospital to commemorate her Majesty's Jubilee? I might mention that our staff was the first to act upon the recommendation contained in Clinical Lectures delivered at St. Bartholomew's Hospital in 1894 to test the above treatment in medical as well as other suitable cases, and the remarkable results obtained in rheumatoid arthritis, acute and chronic gout, rheumatic and neuralgic affections, which have been reported from time to time in the medical press, led to the hiring by this institution of the apparatus necessary to carry the treatment out.

Availing themselves of the occasion of Her Majesty's Jubilee, it was suggested to the proprietors that it would be a graceful way of commemorating the same if they were to present the hospital with the apparatus now on hire. Legal difficulties being in the way of carrying out that suggestion, Mr. Lewis A. Tallerman, recognizing the interest taken in his method of treatment by the staff, personally undertook to defray the cost of placing the apparatus at the disposal of the hospital, conditionally that it should be used for the treatment of the necessitous poor only—*i.e.*, sufferers unable to pay medical fees or to afford donations. To this the committee readily consented, and they desire to gratefully acknowledge Mr. Tallerman's kindness and liberality, which place in their hands a therapeutic measure which has lent such valuable assistance in the past in relieving the pain, suffering and distress amongst a large class of patients visiting the North-West London Hospital.

I am, Sir, yours faithfully,

ALFRED CRASKE, Secretary.

North-West London Hospital, Kentish Town Road.

August 11th.

From THE VORWÄRTS, BERLIN, March 2nd, 1898.

An advance in the Treatment of Articular Rheumatism.

IN the treatment of acute and chronic rheumatism of the joints the numerous methods in vogue chiefly act by means of heat. Naturally this action has limits. Mr. Tallerman, of England, has lately devised an apparatus which permits single limbs to be submitted to the prolonged action of temperatures far above 100° C., and thereby increases the thermal effect in an extraordinary degree. The first specimens of this apparatus which have reached Germany were recently demonstrated at the Medical Congress and on patients. It was proved that an arm or other limb can be subjected to excessively high temperatures for hours together without any discomfort.

Hospital and Medical Reports.

OTHER PRESS NOTICES.

The "TIMES," November 11th, 1895.

"The apparatus has been largely employed at St. Bartholomew's and other hospitals, and the medical officers of those institutions bear testimony to its efficacy in the treatment of various affections. The value of sustained local warmth has long been known, but the means of applying it which have hitherto been in use have been comparatively inefficient."

The "STANDARD," February 1st, 1895.

"A really valuable addition to our present very imperfect means of dealing with a peculiarly trouble-some and unsatisfactory class of cases. . . . The treatment has been applied with striking success to cases of both chronic and acute inflammation of the joints, whether due to gout, rheumatism or injury . . It is clear that the new appliance has a real value, and deserves the attention of the medical profession."

"DAILY CHRONICLE," October 26th, 1895.

. . . The invention seems likely to afford prompt relief, if not indeed actually a complete cure, in cases that remain painful and obstinate under the traditional mode of treatment. In that view it appears to be very important and to demand the careful consideration of medical men."

"DAILY CHRONICLE," July 2nd, 1898.

"The method deserves a thorough trial in the many cases where ordinary medical treatment has proved to be of but little service. . . A really valuable form of aero-therapeutics."

"MORNING LEADER," January 19th, 1897.

"The effect of the treatment is to give the patient in his own house, or in a hospital, more beneficial effects, and in a much more permanent form than those obtained by visiting the various 'hot springs' and other resorts which are accessible only to the rich."

"SOUTH LONDON PRESS," April 4th, 1896.

"A poor woman, whose hand was so affected with rheumatism that she could not shut it or raise it to her head, placed hand and arm in the apparatus. Wonderful to relate the hand became flexible, and the owner used it—for the first time for many months—for the feminine purpose of doing up her hair."

"GLASGOW HERALD," February 19th, 1898.

"The effect of the dry-heat on stiff joints and on limbs, which have been well nigh immovable, is little short of marvellous. The treatment is perfectly consistent with all we know physiologically of the effects of warmth, but it was left for Mr. Tallerman to demonstrate how dry-heat applied by his method could be utilized readily and easily both in private and in hospital practice."

"SCOTSMAN," October 28th, 1895.

"Deserves the attention of medical men as likely to be of use to them in the treatment of an obstinate class of diseases."

"LEEDS MERCURY," June 19th, 1895.

"One of the cases treated yesterday was that of a patient suffering from chronic rheumatism in both wrists. Prior to the treatment every movement of the fingers and hand was painful, and the joints of the wrists were badly swollen. After a forty minutes sitting (in the Tallerman apparatus) the pain had practically ceased, the swellings were considerably reduced, and the patient was able to straighten out her fingers—an utter impossibility before the treatment—and also to grasp with the hand."

"BRADFORD OBSERVER," March 18th, 1898.

"A very valuable addition to the means of treatment of some long-standing and obstinate ailments."

"HARROGATE VISITOR," June 22nd, 1895.

"The Tallerman dry-air bath has more than the reputation of a day behind it, and is an important factor in the treatment of rheumatic affections. The Corporation will do well to add the appliance to the baths under their supervision, as the class of ailments which the hot-air bath benefits is a very common one."

"BUXTON HERALD," February 23rd, 1898.

"Is likely to have an important bearing on the treatment of diseases, the cure of which has so long been associated with the waters and baths of Buxton."

"ABERDEEN FREE PRESS," February 7th, 1898.

"Successfully tried for five years, and that in the hands of able medical men and chiefs of hospitals."

"VANITY FAIR," March 3rd, 1898.

"Mr. Tallerman has very greatly benefited suffering humanity by introducing a treatment for gout, rheumatism and other painful diseases that is always curative and never productive of mischief, or even discomfort."

"CORK HERALD," March 21st, 1896.

"Rapidly making its way to the front as the recognised remedy for the cure of rheumatics, sprains, stiff joints, &c."

"CORK CONSTITUTION," March 20th, 1896.

"Its systematic use shows most beneficial results, numerous cases of chronic rheumatic affections being authentically recorded as cured and relieved."

"CHRISTIAN WORLD," September 3rd, 1896.

"Cases of chronic rheumatism of many years standing, and acute lumbago and sciatica were completely cured by the treatment."

"NEW YORK HERALD," December 15th, 1896.

"Experiments have just been concluded at the Hospital for the Ruptured and Crippled with a new method of treatment for chronic rheumatism and gout, sprains and other diseases of the joints and muscles, which has excited great interest in the medical profession both in England and America. The method is known as the Tallerman treatment by the local application of super-heated dry air. . . . Patients who have been racked with pain for months have been known to fall into a refreshing sleep while undergoing the treatment."

"BOURNEMOUTH OBSERVER," February 26th, 1898.

"There is every reason to believe that in the hands of the medical profession a further investigation into the effects of the external application of dry air at a high temperature, will lead to remarkable results in cases which have hitherto been regarded as beyond the influence of such an agency."

"SCOTSMAN," January 28th, 1898.

"An ingenious and humane invention that has already done much now promises to do more for the alleviation of a too common pain."

"LEICESTER POST," February 19th, 1898.

"Deserves the special and sympathetic attention of all who desire to promote the cure of a group of more or less painful diseases."

"WESTERN MORNING NEWS," February 11th, 1898.

"The results obtained by this new treatment should materially assist those who suffer from what has been called the 'national disease of England.'"

"YORKSHIRE DAILY POST," February 2nd, 1898.

"The treatment has been so widely tested in hospitals, as well as in private practice, that it must now emerge from the region in which scepticism reigns into one of general approval."